Math Challenge I-C
Topics in Algebra

Areteem Institute

Math Challenge I-C Topics in Algebra

Edited by   John Lensmire
             David Reynoso
             Kevin Wang
             Kelly Ren

PUBLISHED BY ARETEEM PRESS

ISBN: 1-944863-17-6
ISBN-13: 978-1-944863-17-3
First printing, August 2018.

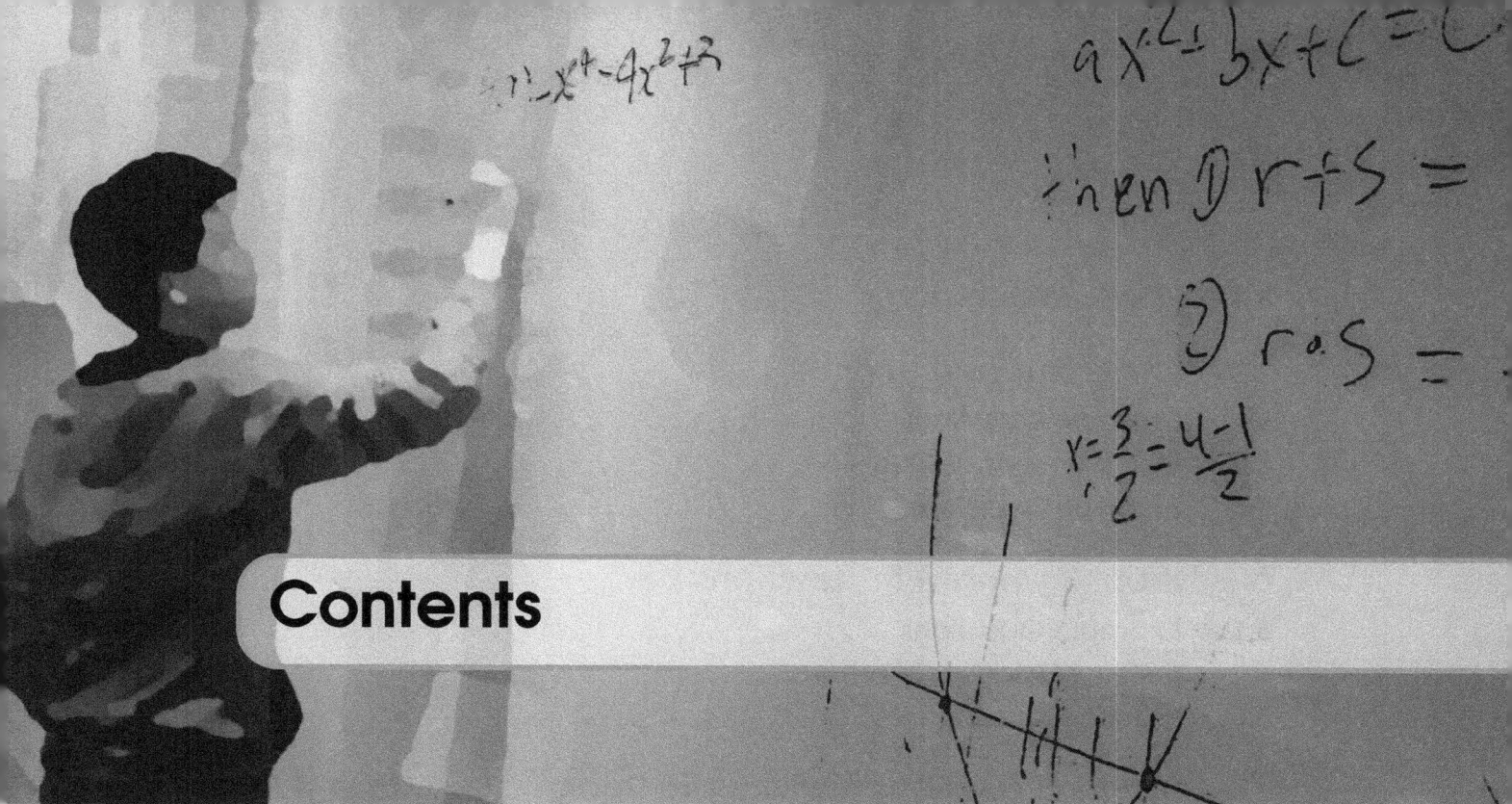

Contents

Introduction

The math challenge curriculum textbook series is designed to help students learn the fundamental mathematical concepts and practice their in-depth problem solving skills with selected exercise problems. Ideally, these textbooks are used together with Areteem Institute's corresponding courses, either taken as live classes or as self-paced classes. According to the experience levels of the students in mathematics, the following courses are offered:

- Fun Math Problem Solving for Elementary School (grades 3-5)
- Algebra Readiness (grade 5; preparing for middle school)
- Math Challenge I-A Series (grades 6-8; intro to problem solving)
- Math Challenge I-B Series (grades 6-8; intro to math contests e.g. AMC 8, ZIML Div M)
- Math Challenge I-C Series (grades 6-8; topics bridging middle and high schools)
- Math Challenge II-A Series (grades 9+ or younger students preparing for AMC 10)
- Math Challenge II-B Series (grades 9+ or younger students preparing for AMC 12)
- Math Challenge III Series (preparing for AIME, ZIML Varsity, or equivalent contests)
- Math Challenge IV Series (Math Olympiad level problem solving)

These courses are designed and developed by educational experts and industry professionals to bring real world applications into the STEM education. These programs are ideal for students who wish to win in Math Competitions (AMC, AIME, USAMO, IMO,

ARML, MathCounts, Math League, Math Olympiad, ZIML, etc.), Science Fairs (County Science Fairs, State Science Fairs, national programs like Intel Science and Engineering Fair, etc.) and Science Olympiad, or purely want to enrich their academic lives by taking more challenges and developing outstanding analytical, logical thinking and creative problem solving skills.

Math Challenge I-C is a four-part course designed to bridge the middle school and high school math materials. For students who participate in the American Math Competitions (AMC), there is a big gap in both the fundamental math concepts and the problem-solving techniques involved between the AMC 8 and AMC 10 contests. This course is developed to help students transition smoothly from middle school to high school, and prepare them for high school math competitions including the AMC 10 & 12, ARML, and ZIML. The full course covers topics and introductory problem solving in algebra, geometry, and finite math. Algebraic topics include linear equations, systems of equations and inequalities, exponents and radicals, factoring polynomials, and solving quadratic equations. Geometric topics include angles in triangles, quadrilaterals, and polygons, congruent and similar polygons, calculating area, and algebraic geometry. Topics in finite math include logic, introductory number theory, and an introduction to probability and statistics. These topics serve as the fundamental knowledge needed for a more advanced problem solving course such as Math Challenge II-A.

The course is divided into four terms:

- Summer, covering Algebra
- Fall, covering covering additional topics in Algebra
- Winter, covering Geometry
- Spring, covering Finite Math

The book contains course materials for Math Challenge I-C: Additional topics in Algebra.

We recommend that students take all four terms starting with the Summer, but students with the required background are welcome to join for later terms in the course.

Students can sign up for the course at `classes.areteem.org` for the live online version or at `edurila.com` for the self-paced version.

About Areteem Institute

Areteem Institute is an educational institution that develops and provides in-depth and advanced math and science programs for K-12 (Elementary School, Middle School, and High School) students and teachers. Areteem programs are accredited supplementary programs by the Western Association of Schools and Colleges (WASC). Students may attend the Areteem Institute in one or more of the following options:

- Live and real-time face-to-face online classes with audio, video, interactive online whiteboard, and text chatting capabilities;
- Self-paced classes by watching the recordings of the live classes;
- Short video courses for trending math, science, technology, engineering, English, and social studies topics;
- Summer Intensive Camps held on prestigious university campuses and Winter Boot Camps;
- Practice with selected free daily problems and monthly ZIML competitions at ziml.areteem.org.

Areteem courses are designed and developed by educational experts and industry professionals to bring real world applications into STEM education. The programs are ideal for students who wish to build their mathematical strength in order to excel academically and eventually win in Math Competitions (AMC, AIME, USAMO, IMO, ARML, MathCounts, Math Olympiad, ZIML, and other math leagues and tournaments, etc.), Science Fairs (County Science Fairs, State Science Fairs, national programs like Intel Science and Engineering Fair, etc.) and Science Olympiads, or for students who purely want to enrich their academic lives by taking more challenging courses and developing outstanding analytical, logical, and creative problem solving skills.

Since 2004 Areteem Institute has been teaching with methodology that is highly promoted by the new Common Core State Standards: stressing the conceptual level understanding of the math concepts, problem solving techniques, and solving problems with real world applications. With the guidance from experienced and passionate professors, students are motivated to explore concepts deeper by identifying an interesting problem, researching it, analyzing it, and using a critical thinking approach to come up with multiple solutions.

Thousands of math students who have been trained at Areteem have achieved top honors and earned top awards in major national and international math competitions, including Gold Medalists in the International Math Olympiad (IMO), top winners and qualifiers at the USA Math Olympiad (USAMO/JMO) and AIME, top winners at the

Zoom International Math League (ZIML), and top winners at the MathCounts National Competition. Many Areteem Alumni have graduated from high school and gone on to enter their dream colleges such as MIT, Cal Tech, Harvard, Stanford, Yale, Princeton, U Penn, Harvey Mudd College, UC Berkeley, or UCLA. Those who have graduated from colleges are now playing important roles in their fields of endeavor.

Further information about Areteem Institute, as well as updates and errata of this book, can be found online at `http://www.areteem.org`.

Acknowledgments

This book contains many years of collaborative work by the staff of Areteem Institute. This book could not have existed without their efforts. Huge thanks go to the Areteem staff for their contributions!

The examples and problems in this book were either created by the Areteem staff or adapted from various sources, including other books and online resources. Especially, some good problems from previous math competitions and contests such as AMC, AIME, ARML, MATHCOUNTS, and ZIML are chosen as examples to illustrate concepts or problem-solving techniques. The original resources are credited whenever possible. However, it is not practical to list all such resources. We extend our gratitude to the original authors of all these resources.

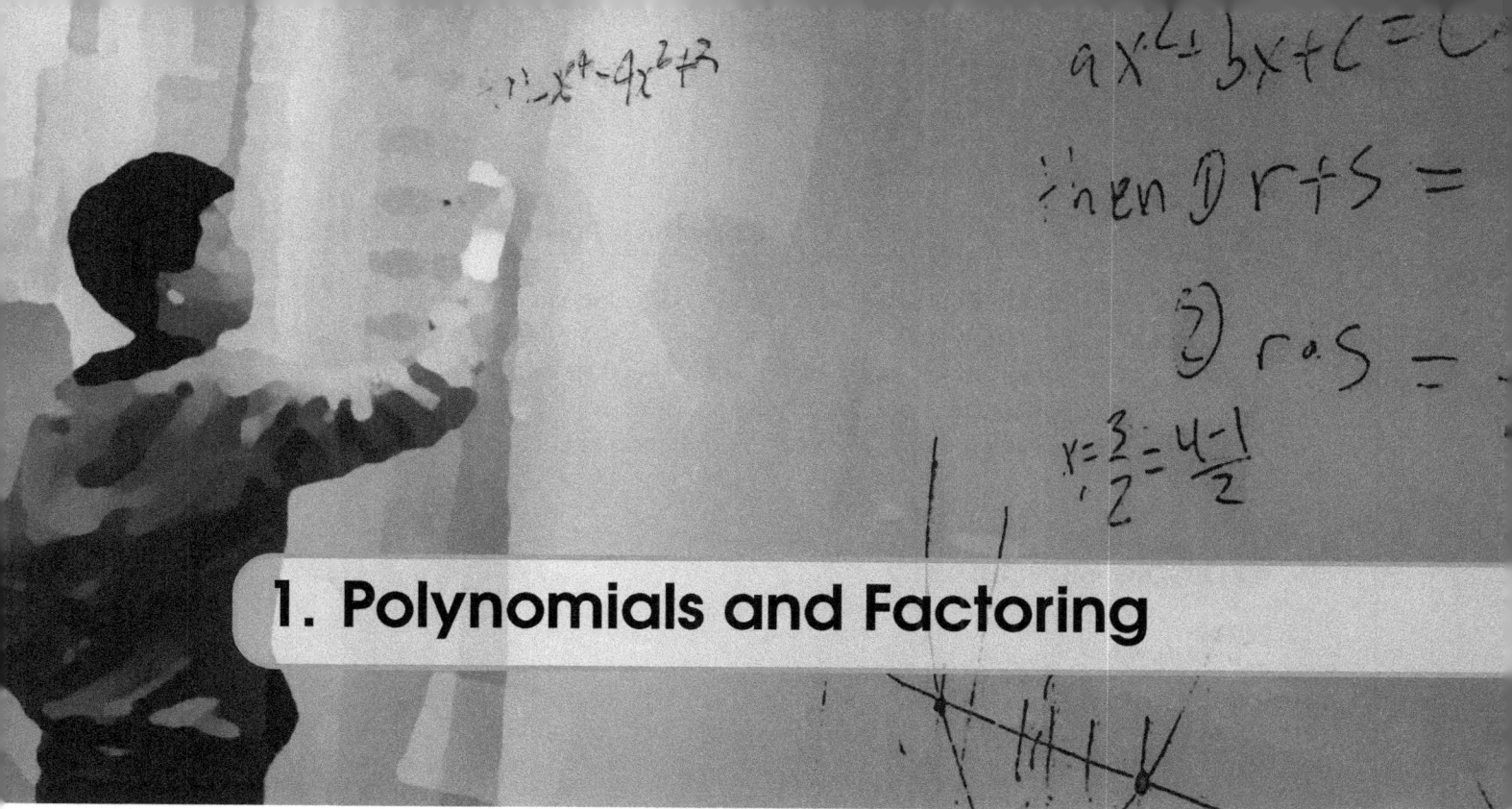

1. Polynomials and Factoring

Polynomials

- A *polynomial* is an expression that is made of adding multiples of powers of a variable or variables.
  - For example, $5x^3 - 2x + 1$, $-a^{99} - 9a^9 - 99$, $4z^2 - 12z + 9$, and $9x^5y^5 + 20x^3y^4$ are all polynomials.
  - Usually polynomials are written with the powers going up or going down.
- A *monomial* or *term* refers to a polynomial with exactly one piece.
  - Every polynomial can be thought of as a sum of monomials / terms. For example, the polynomial $x^7 - 4x^5 + 18x$ is the sum of the monomials x^7, $-4x^5$, and $18x$.
- The *coefficient of a term* is the number being multiplied by a power.
  - The coefficient of x^3 in $5x^3 - 2x + 1$ is 5, the coefficient of a^9 in $-a^{99} - 9a^9 - 99$ is -9, and the coefficient of z^7 in $4z^2 - 12z + 9$ is 0.
- The *degree of a polynomial* is the largest power in a polynomial.
  - The degrees of the polynomials $5x^3 - 2x + 1$, $-a^{99} - 9a^9 - 99$, and $4z^2 - 12z + 9$ are 3, 99, and 2.
  - A constant is also a polynomial, and it has degree 0 if the constant itself is nonzero.
- It is often useful to write a polynomial as the product of other polynomials. This is called *factoring* a polynomial.

Zero Factor Property

- If we want to solve the equation $A \times B = 0$ this is the same as solving the compound equation $A = 0$ or $B = 0$.
  - For example if we want to solve $x(x-1) = 0$ we either have $x = 0$ or $x - 1 = 0$ so $x = 0$ or $x = 1$ are the solutions.
  - Using this property we can solve polynomial equations using factoring.

1.1　Example Questions

Problem 1.1 Review of Distributive Property

(a) Expand $2x(x^2 + 3)$.

(b) Expand $-3xy(x^3 + 2xy + y^2)$.

Problem 1.2 Factor the following by pulling out common terms.

(a) $2x^3 - 4x$.

(b) $12x^4 + 3x^3 + 6x^2$.

Problem 1.3 Review of expanding binomials (FOIL)

(a) Expand $(x+3)(x+4)$.

(b) Expand $(2x+3)(x-1)$.

(c) Expand $(3x^2 + y)(y + 2)$.

Problem 1.4 Factor the following.

(a) $x^2 + 5x + 6$.

(b) $x^2 - x - 12$.

(c) $2x^2 - 7x + 3$.

Problem 1.5 Solve the following equations

(a) $x^2 + 2x = 0$.

(b) $x^2 - 3x + 2 = 0$.

(c) $2x^2 - 15x - 27 = 0$.

Problem 1.6 Albert was playing baseball and hit a ball straight up with an initial speed of 47 feet per second. The height of the ball was modeled by the equation $h = -16t^2 + 47t + 3$. How long did it take for the ball to come back down and hit the ground?

Problem 1.7 Solve the following.

(a) $y(y-6)+y(y+1)+25 = 2y(y+5)-5$.

(b) $2y(y+3) = 8$.

Problem 1.8 Practice with Polynomials

(a) Calculate $2x(x^2+3x+2)-3x^2(x+2)$.

(b) Expand $(x^2+2)(x^2-3x+2)$.

Problem 1.9 Solve the following polynomial equations.

(a) $-2x^3-2x^2+12x = 0$

(b) $2x^2(x-1)+9x(x-1)-5x+5 = 0$.

Problem 1.10 Hank is planning building a jacuzzi in his backyard. He initially planned to make the jacuzzi 4 feet by 6 feet. After thinking some more, he decided he wanted to double the area of his jacuzzi. If he still wants the jacuzzi to be 2 feet longer than it is wide, how many feet should he add to the dimensions of the jacuzzi to double the area?

1.2 **Quick Response Questions**

Problem 1.11 What is the degree of the polynomial $2x^2 + 5x^4 + 3x^3 + x$?

Problem 1.12 What is the coefficient of x^3 in the polynomial $2x^2 + 5x^4 + 3x^3 + x$?

Problem 1.13 Consider polynomials $5x^3 - 2$ and $x^2 + 2$. What is the degree of $(5x^3 - 2) + (x^2 + 2)$?

Problem 1.14 Consider polynomials $5x^3 - 2$ and $x^2 + 2$. What is the degree of $(5x^3 - 2)(x^2 + 2)$?

Problem 1.15 Expanding $(x + 5)(x - 6)$ gives

(A) $x^2 + 11x + 30$
(B) $x^2 - x + 30$
(C) $x^2 - x - 30$
(D) $x^2 + x - 30$

Problem 1.16 Which is the correct factoring of $x^2 + 2x - 8$?

(A) $(x + 4)(x - 2)$
(B) $(x - 4)(x + 2)$
(C) $(x + 4)(x + 2)$
(D) $(x - 4)(x - 2)$

Problem 1.17 What are the solutions to $4x = 2x^2$?

(A)   only $x = 2$

(B)   only $x = 0$

(C) $x = 2$ or $x = 4$

(D) $x = 0$ or $x = 2$

Problem 1.18 The equation $3x^2 + 10x - 8 = 0$ has one positive and one negative solution. What is the positive solution? Round your answer to the nearest hundredth if necessary.

Problem 1.19 Is $x = -2$ a solution to $x^3 + 3x^2 + 3x + 2 = 0$?

Problem 1.20 How many integer solutions does $2x^3 + 3x^2 + x = 0$ have?

1.3 Practice Questions

Problem 1.21 Expand using the Distributive Property

(a) $3x(x + 4x^2)$

(b) $5y^3(3x^2y + 6xy + 9y^2)$

Problem 1.22 Expand the following binomials

(a) $(4x - 2)(x + 4)$

(b) $(x + 4y)(3y - x)$

Problem 1.23 Factor the following polynomials of degree 2

(a) $x^2 + 8x + 12$

(b) $3x^2 - 15x + 18$

Problem 1.24 Solve the following equations

(a) $4x^2 - 8x - 60 = 0$

(b) $5x^2 + 10x - 40 = 0$

Problem 1.25 Solve the following equations

(a) $x(x-4) + x(x+2) - 8 = 8(9-x)$

(b) $y(y+2) - 2y^2 + 24 = 5y(10-y) - 20$

Problem 1.26 Dylan drops a ball from the top of a building that is 64 feet tall. The height of the ball in feet after t seconds can be modeled by the equation $h = 64 - 16t^2$. How many seconds does it take for the ball to hit the ground?

Problem 1.27 Calculate

(a) $5x(8x^2 - 8x + 8) + 8(5x^2 + 5x - 5)$

(b) $(y^2 + 3)(2y^3 + 5y - 4)$

Problem 1.28 Solve the equation $2x^3 + 14x^2 - 60x = 0$.

Problem 1.29 Greta needs to build a fence around her rectangular yard. The fence will be three times as long as it is wide and will covers an area of 48 square meters. How many meters of fence does Greta need for her yard?

Problem 1.30 Audra wants to build an open squared carboard box of height 5 inches that can hold a volume of 245 cubic inches. She will start with a square piece of cardboard and will cut squares of side 5 inches on each of the corners. She will then fold the piece she is left with to form the box. What is the length of the side of the piece of cardboard before she cuts the corners?

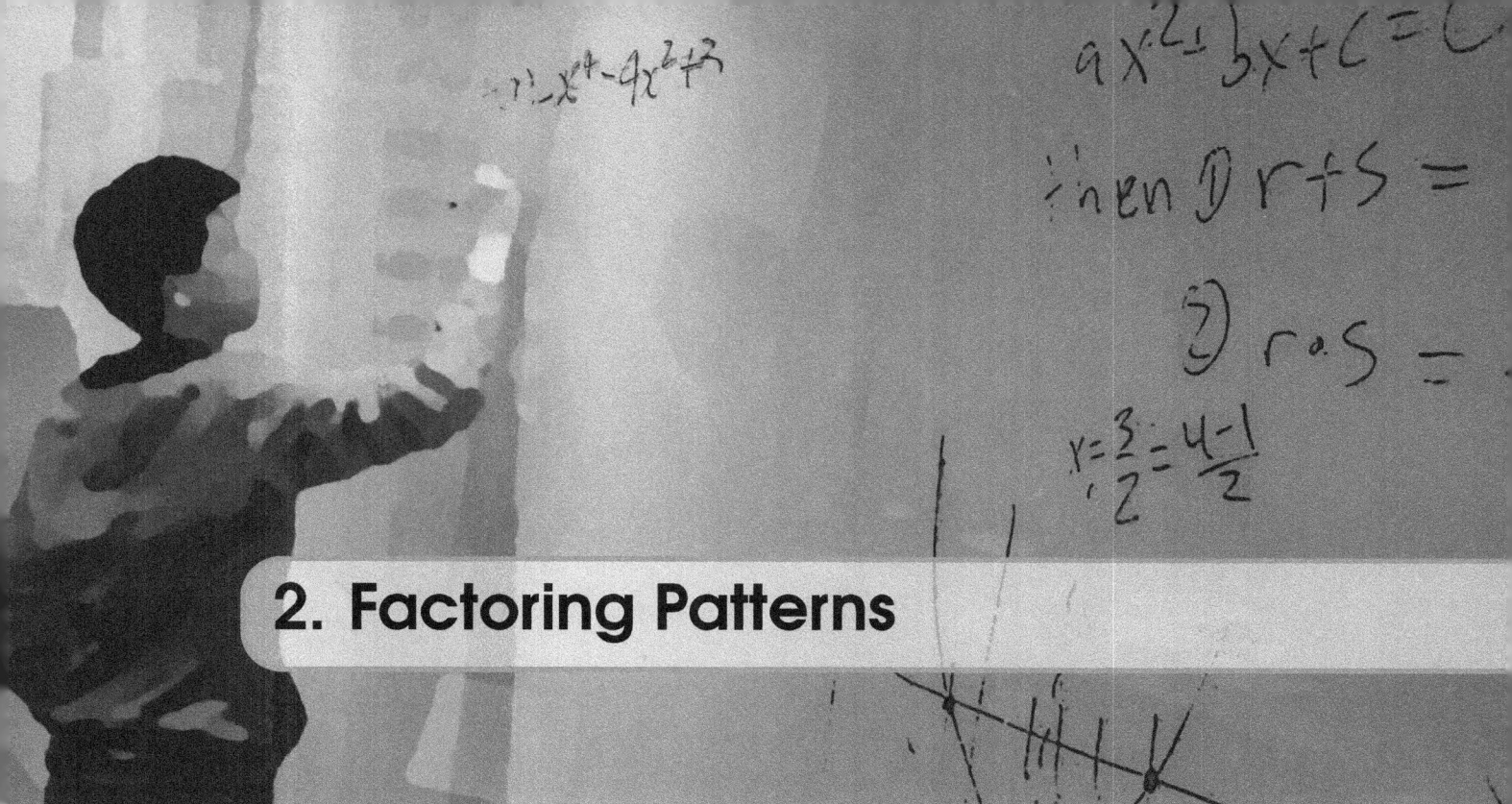

2. Factoring Patterns

Review of Factoring

- Recall factoring a polynomial is writing it as a product of other polynomials.
- Looking for common factors is a good first step for factoring.
- Tricks such as Reverse FOIL or patterns are also useful.
- We'll learn some other patterns for factoring today.

Factor Theorem

- Suppose you are solving a polynomial equation $p(x) = 0$ where $p(x)$ is a polynomial. Then the following two statements are equivalent:
    - $x = r$ is a solution to the equation $p(x) = 0$.
    - $(x - r)$ is a factor of the polynomial $p(x)$.

Summary of Factoring Patterns

- Powers of $(a + b)$ and Pascal's Triangle
    - $(a+b)^2 = a^2 + 2ab + b^2$
    - $(a+b)^3 = a^3 + 3a^2b + 3ab^2 + b^3$
    - $(a+b)^4 = a^4 + 4a^3b + 6a^2b^2 + 4ab^3 + b^4$
- Difference of two squares
    - $(a-b)(a+b) = a^2 - b^2$
- Difference of two cubes
    - $(a-b)(a^2 + ab + b^2) = a^3 - b^3$
- Sum of two cubes

$$- (a+b)(a^2 - ab + b^2) = a^3 + b^3$$

2.1 Example Questions

Problem 2.1 Finding our first pattern. Expand the following:

(a) $(a+b)^2$

(b) $(a+b)^3$

(c) $(a+b)^4$

Problem 2.2 Review of Factoring.

(a) Factor $x^2 - 3x - 4$.

(b) Factor $2x^2 + 8x + 8$.

(c) Factor $x^2 - 9$

Problem 2.3 Solve the following equations.

(a) $x^2 - x + \dfrac{1}{4} = 0$.

(b) Solve $x^4 - 16 = 0$.

Problem 2.4 Solve the following equations involving exponents.

(a) $x^2 = 25$.

(b) $x^3 = -27$.

(c) $x^4 = -64$.

Problem 2.5 Expand the following.

(a) $(x+3)(x^2 - 3x + 9)$

(b) $(a+b)(a^2 - ab + b^2)$

(c) $(a-b)(a^2 + ab + b^2)$

Problem 2.6 Factor Theorem Practice

(a) Suppose $(x-2)(x+k) = 0$ has solutions of 2 and 3. What is k?

(b) Is $(x-2)$ a factor of $x^4 - 2x^3 + 2x^2 - 8x + 8$?

Problem 2.7 Practice the Factoring Patterns!

(a) Factor $8x^3 + 125$.

(b) $x^8 - 1$

(c) $8x^3 + 12x^2 + 6x + 1$

Problem 2.8 Solve the following equations.

(a) $x^4 - 6x^2 + 9 = 0$.

(b) $(x+2)^2 = 8$.

Problem 2.9 Factoring Tricks

(a) Factor $x^2 - 1 + x^3 - 1$.

(b) Factor $x^3 + x^2 + x + 1$.

Problem 2.10 Solving Equations Tricks

(a) Solve the equation $x^2 - 6x + 9 = 5$.

(b) Solve the equation $xy + x - 2y - 2 = 0$.

2.2 Quick Response Questions

Problem 2.11 Expand $(2x-1)^2$ to get $Ax^2 + Bx + C$ for integers A, B, C. What is B?

Problem 2.12 Expand $(x-1)^3$.

(A) $x^3 - 1$
(B) $x^3 + 3x^2 + 3x + 1$
(C) $x^3 - 3x^2 + 3x - 1$
(D) $x^3 - 3x^2 - 3x - 1$

Problem 2.13 Expand $(a+b)^5$. What is the coefficient of a^3b^2?

Problem 2.14 Let K be a real number and N be a positive even integer. Is it true that $x^N = K$ has 2 solutions?

Problem 2.15 Let $p(x) = x^2 + bx + c$, so $p(x)$ is a quadratic. Suppose the two solutions to $p(x) = 0$ are 3 and -4. What is the factored form of $p(x)$?

(A) $(x-3)(x-3)$
(B) $(x-3)(x+4)$
(C) $(x+3)(x-4)$
(D) $(x+3)(x+4)$

Problem 2.16 The equation $-3x^3 + 18x^2 - 36x + 24 = 0$ has one integer solution. What is this solution?

Problem 2.17 What is the sum of the solutions to the equation $(x-3)^2 = 6$? Round your answer to the nearest tenth if necessary.

Problem 2.18 Which of the following is not a solution to $x^2 - 4y^2 = 0$?

(A) $x = 4, y = 2$
(B) $x = -2, y = -1$
(C) $x = 1, y = -2$
(D) $x = 2\sqrt{3}, y = -\sqrt{3}$

Problem 2.19 Factor $xy + xz + yw + wz$.

(A) $(x+w)(y-z)$
(B) $(x-y)(w-z)$
(C) $(x+y)(w+z)$
(D) $(x+w)(y+z)$

Problem 2.20 Which of the following is NOT a factored form of $x^6 - 1$?

(A) $(x^3 - 1)(x^3 + 1)$
(B) $(x+1)(x^5 + x^4 + x^3 + x^2 + x + 1)$
(C) $(x^2 - 1)(x^4 + x^2 + 1)$
(D) $(x-1)(x+1)(x^4 + x^2 + 1)$

2.3 Practice Questions

Problem 2.21 Expand the following
(a) $(x+3)^2$

(b) $(x-4)^3$

Problem 2.22 Factor $x^2 - 256$

Problem 2.23 Solve for x in the equation $x^2 - \dfrac{x}{2} + \dfrac{1}{16} = 0$

Problem 2.24 Solve the equation $(x-2)^2 = 81$

Problem 2.25 Verify if the following are factors of $x^4 + 7x^3 + 8x^2 - 28x - 48$
(a) $(x+4)$

(b) $(x-1)$

Problem 2.26 Factor the expression $27x^3 + 343$

Problem 2.27 Solve the equation $4x^2 - 49 = 0$

Problem 2.28 Solve the following equation $x^2 + 8x + 16 = 625$.

Problem 2.29 Factor the expression by grouping $x^3 + 3x^2 + 2x + 6$

Problem 2.30 Solve the equation $xy - 3x + 4y - 12 = 0$ for x and y.

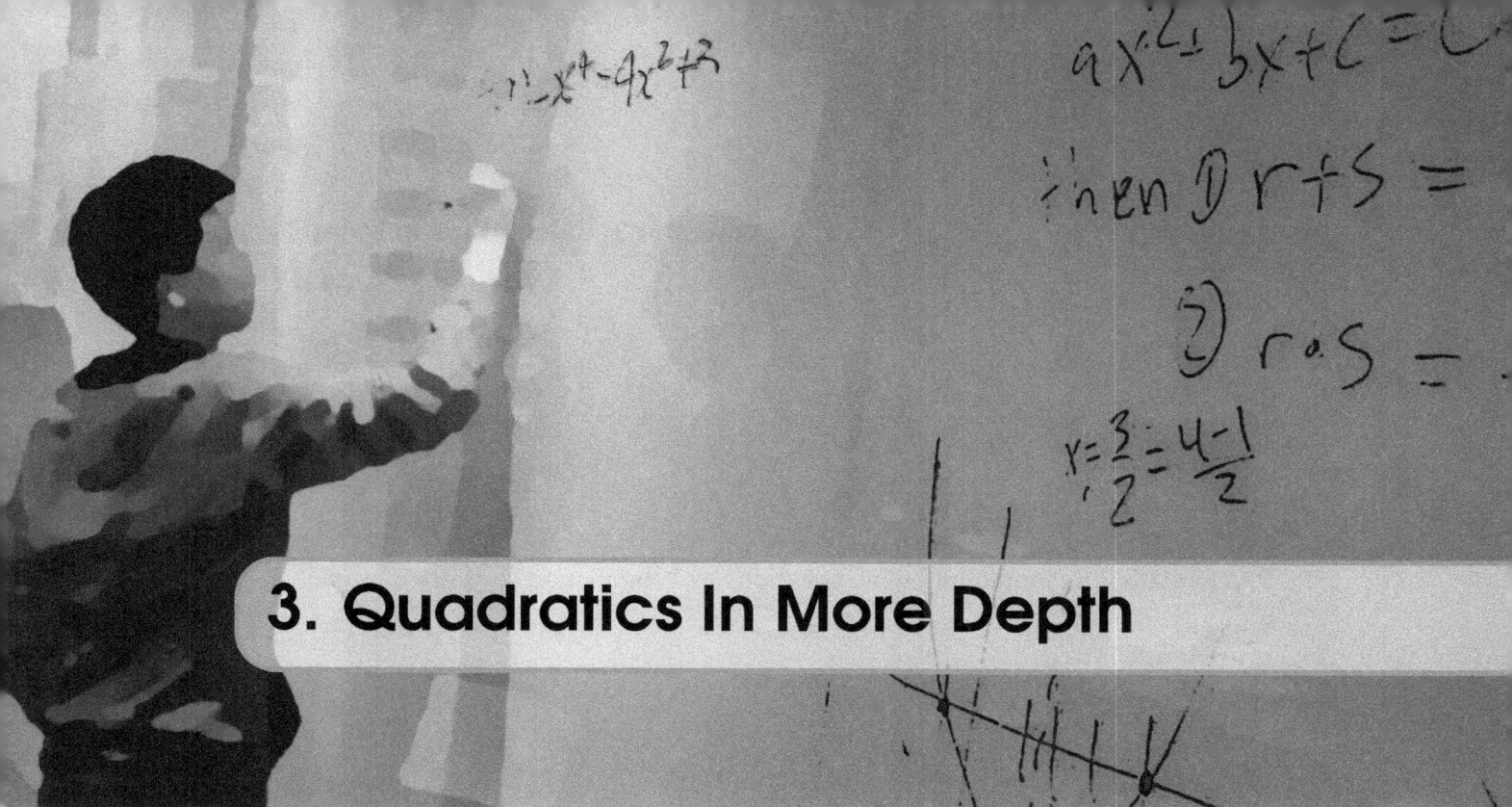

3. Quadratics In More Depth

Quadratic Equations

- Recall quadratic equations are of the form $ax^2 + bx + c = 0$.
- In general, quadratic equations have up to 2 real solutions.
    - We have seen factoring is one way to solve quadratic equations.
    - Today we will explain a general method for solving any quadratic equation.

Graphing Quadratic Equations

- The graph of an equation of the form $y = ax^2 + bx + c$ is called a parabola.
- The general shape of a parabola is either an upwards or downwards "U".
- Points of interest are:
    - x-intercepts, where the parabola crosses the x-axis. These occur when $y = 0$.
    - y-intercept, where the parabola crosses the y-axis. This occurs when $x = 0$.
    - The vertex, which is the bottom or top of the "U". The is where the y value is largest (called the maximum) or smallest (called the minimum).

3.1 Example Questions

Problem 3.1 Review of Solving Quadratic Equations

(a) $x^2 + 6x + 8 = 0$

(b) $x^2 + 6x + 9 = 3$

Problem 3.2 Solve the following equations by "completing the square".

(a) $x^2 - 4x - 5 = 0$

(b) $x^2 - 4x - 1 = 0$

Problem 3.3 Solve the equation $2x^2 + 9x + 3 = 0$ by completing the square.

Problem 3.4 Solve the equation $ax^2 + bx + c = 0$ by completing the square.

Problem 3.5 Find solutions to the following using the quadratic formula.

(a) $3x^2 + 4x - 4 = 0$

(b) $x^2 - 6x + 2 = 0$

Problem 3.6 Graph the equation $y = x^2 + 6x + 5$. Label the vertex, x-intercepts, and y-intercept.

Problem 3.7 Graph the equation $y = a(x-h)^2 + k$

Problem 3.8 Graph the equation $y = -2x^2 + x - 5$. Label the vertex and any intercepts. Is the vertex a maximum or minimum?

Problem 3.9 For the equation $y = 4x - x^2$ find the intercepts and vertex. Is the vertex a maximum or minimum?

Problem 3.10 During their whale watching tour, Jason took a video of a dolphin jumping out of the water. He found out he could model the height of the dolphin during its jump using the quadratic equation $j(t) = 32t - 16t^2$. What was the maximum height of the dolphin during the jump?

3.2 Quick Response Questions

Problem 3.11 When does a parabola intercept the x-axis?

(A) When $x = y$
(B) When $x = 0$
(C) When $y = 0$
(D) None of the above

Problem 3.12 When does a parabola intercept the y-axis?

(A) When $x = y$
(B) When $x = 0$
(C) When $y = 0$
(D) None of the above

Problem 3.13 What is the y-intercept of the parabola $y = 2x^2 + 4x + 3$?

Problem 3.14 Does the parabola $y = 4x^2 + 5x - 9$ open upwards?

Problem 3.15 Does the parabola $y = -2(x+4)^2 + 8$ open upwards?

Problem 3.16 What is the y-intercept of the parabola $y = (x-4)^2 + 4$

Problem 3.17 The parabola $y = (x-4)^2$ has one x-intercept, what is it?

Problem 3.18 What is the vertex of the parabola $y = (x+4)^2 + 8$?

(A) $(4,8)$
(B) $(-4,-8)$
(C) $(4,-8)$
(D) $(-4,8)$

Problem 3.19 What is the vertex of the parabola $y = x^2 + 4x + 6$?

(A) $(4,6)$
(B) $(-2,2)$
(C) $(4,-6)$
(D) None of the above

Problem 3.20 Does the parabola $y = -4x^2 + 2x - 9$ have a maximum?

3.3 Practice Questions

Problem 3.21 Quadratic Equations Review
(a) $x^2 + 6x - 16 = 0$

(b) $x^2 - 7x + 10 = 0$

Problem 3.22 Complete the square in $x^2 + 8x - 32$

Problem 3.23 Solve the equation by completing the square $16x^2 - 96x + 139 = 0$

Problem 3.24 Solve the equation $ax^2 + c = 0$. Be sure to simplify your answer.

Problem 3.25 Solve the following using the quadratic formula.

(a) $x^2 - 2x - 1 = 0$

(b) $5x^2 - 20x + 5 = 0$

Problem 3.26 Graph the equation $y = x^2 - 3$. Label the vertex, x-intercepts, and y-intercept.

Problem 3.27 Graph the equation $y = 2x^2 + 16x + 28$ and label the vertex and intercepts.

Problem 3.28 For the equation $y = 12 - x^2 - 4x$ find the intercepts and the vertex. Is the vertex a maximum or a minimum?

Problem 3.29 Julius will collect fund for his school with a cookie sale. He figured out that if he sells a box of cookies at x dollars per box, then he will get a total profit of $P = -(x - 4)^2 + 12$ dollars. What is the price per box that will let Julius get the maximum possible profit? What is the maximum possible profit?

Problem 3.30 After Julius's sucess with the cookie sale, his friend Randle decided to try selling cookies for himself. He was more worried in the cost per box it would cost him to buy the cookies before the sale. Randle figured out that the cost per box could be modeled using the equation $C = x^2 - 40x + 403$, where x is the number of boxes bought. What is the minimum price Randle could pay for a box of cookies? How many boxes would he need to buy to get this price?

4. Prob. Solving with Quadratics

Review of Quadratic Equations

- Last handout we covered a general formula, the *quadratic formula*, for solving quadratic equations. The solutions to the equation $ax^2 + bx + c = 0$ are $\dfrac{-b \pm \sqrt{b^2 - 4ac}}{2a}$.

Review of Graphing Quadratic Equations

- The graph of an equation of the form $y = ax^2 + bx + c$ is called a parabola.
- The general shape of a parabola is either an upwards or downwards "U".
- Points of interest are:
    - x-intercepts, where the parabola crosses the x-axis. These occur when $y = 0$.
    - y-intercept, where the parabola crosses the y-axis. This occurs when $x = 0$.
    - The vertex, which is the bottom or top of the "U". The is where the y value is largest (called the maximum) or smallest (called the minimum).

Symmetry in Parabolas

- If the vertex of a parabola is (h, k), the line $x = h$ is a line of symmetry for the parabola.
- In particular, this means that if the parabola has x-intercepts, that the x-coordinate of the vertex is the average of the x-intercepts.

The Discriminant

- In a quadratic equation $ax^2 + bx + c = 0$ we call the number $b^2 - 4ac$ the discriminant:
    - If the discriminant is positive, then the equation has 2 solutions.
    - If the discriminant is 0, then the equation has 1 solution.
    - if the discriminant is negative, then the equation has 0 solutions.

4.1　Example Questions

Problem 4.1 Practice and Review using the Quadratic Formula

(a) Solve the equation $x^2 + 4x - 21 = 0$.

(b) Solve the equation $2x^2 - 2x - 5 = 0$.

Problem 4.2 Graph the equation $y = x^2 - 2x$. Be sure to label the vertex and any intercepts.

Problem 4.3 Consider the graph of the equaton $y = x^2 - 2x$.

(a) Do you notice any symmetry in the graph? How does this symmetry relate to the x-intercepts and the vertex?

(b) More formally, we would say that $x = 1$ is a line of symmetry for the graph $y = x^2 - 2x$. This means that the y value for $x = 1 + k$ is the same as the y value for $x = 1 - k$ for any real number k. Verify this statement.

Problem 4.4 Consider general graphs of parabolas (so equations of the form $y = ax^2 + bx + c$).

(a) Do all parabolas have two x-intercepts? If not, what are the possibilities for the number of x-intercepts?

(b) The parabola $y = ax^2 + bx + c$ has "x-intercepts" of $\dfrac{-b \pm \sqrt{b^2 - 4ac}}{2a}$. What part of this equation leads to the case where there are 1 or 0 x-intercepts?

Problem 4.5 How many solutions do each of the following equations have? Use the discriminant.

(a) $x^2 + 3x + 2 = 0$.

(b) $7x^2 + 14x + 8 = 0$.

(c) $18x^2 - 21x - 7 = 0$.

Problem 4.6 Dylan wants to throw a ball straight up in the air with a height (in feet) given by $h = -16t^2 + 35t + 5$. His friend Bob is filming horizontally out of his apartment's balcony at a height of 20 feet. How many times will the ball appear on the camera?

Problem 4.7 Carrie has an 8 by 10 inch photo she wants to have framed. The photo should fit evenly in the frame (same amount of space on each side), and she wants the area of the frame to be 150% the area of the photo. How big of a frame should Carrie buy?

Problem 4.8 Tony just got a new house, and had to spend most of his savings on a down payment. After he bought the house, he also had to spend more of his savings for moving, new furniture, etc. He modeled the balance of his savings account using the equation $S = \frac{1}{2}w^2 - 10x + 200$ where S is the amount in his account and w is the number of weeks after he bought he house. Does Tony run out of money?

Problem 4.9 A rectangle has a perimeter of 20 inches. What is the maximum area that this rectangle could have?

Problem 4.10 Consider the equation $3x^2 + bx + 12 = 0$. For what values of b does this equation have 2 real solutions?

4.2   Quick Response Questions

Problem 4.11 How many x-intercepts does the parabola $y = x^2 + 5$ have?

Problem 4.12 What is the minimum y value for the parabola $y = x^2 + 5$?

Problem 4.13 When does a quadratic equation have 2 solutions?

(A)   When the discriminant is negative
(B)   When the discriminant is zero
(C)   When the discriminant is positive
(D)   None of the above

Problem 4.14 What is the discriminant of $2x^2 + 3x - 5 = 0$?

Problem 4.15 How many solutions does the equation $x^2 + 4x - 3 = 0$ have?

Problem 4.16 How many solutions does the equation $4x^2 - 12x + 9 = 0$ have?

Problem 4.17 The parabola $y = x^2 - 8x + 4$ has vertex $(4, -12)$. What is the equation of its line of symmetry?

(A) $x = 4$
(B) $y = 4$
(C) $x = -12$
(D) $y = -12$

Problem 4.18 The parabola $y = x^2 - 18x + 72$ intersects the x-axis at $(6, 0)$ and $(12, 0)$. What is the equation of its line of symmetry?

(A) $x = 6$
(B) $x = 9$
(C) $y = 12$
(D) $y = 9$

Problem 4.19 The parabola $y = x^2 - 18x + 72$ intersects the x-axis at $(6, 0)$ and $(12, 0)$. Which of the following is true about its minimum or maximum?

(A) It has a minimum value of 72 when x=0.
(B) It has a minimum value of -9 when x=9.
(C) It has a maximum value of 72 when x=0.
(D) It has a maximum value of -9 when x=9.

Problem 4.20 A parabola has vertex $(3, -2)$ and contains the point $(6, -7)$. What is the y-intercept of the parabola?

4.3 Practice Questions

Problem 4.21 Practice using the Quadratic Formula
(a) Solve the equation $3x^2 + 2x - 8 = 0$.

(b) Solve the equation $x^2 + 8x - 12 = 0$.

Problem 4.22 Graph the equation $y = x^2 + 4x + 5$. Be sure to label the vertex and any intercepts.

Problem 4.23 Graph the equation $y = x^2 + 3x - 4$. Be sure to label the vertex and any intercepts.

Problem 4.24 Suppose you have a parabola with line of symmetry $x = 8$. The points $(2, -24)$ and $(10, -56)$ are both on the parabola. Find the coordinates of another point that is also on the parabola.

Problem 4.25 Use the discriminant to figure out how many real solutions each of the following equations have:
(a) $3x^2 - 4x + 9 = 0$

(b) $2x^2 - 8x = 12$

Problem 4.26 For what values of m does the equation $x^2 + mx + 4 = 0$ have exactly one solution?

Problem 4.27 For what values of m does the equation $x^2 + 3x - m + 6$ have no solution?

Problem 4.28 Julia wants to fence a rectangular garden in her backyard. Whe wants the length of the garden to be three times the width, and she also wants the garden to cover an area of 75 square yards. Set up a quadratic equation and figure out what are the dimensions of her garden.

Problem 4.29 Troy shoots an arrow while his friends record it from afar. After looking at the recording, they noticed that the arrow followed the path of a parabola. If it took 2.3 seconds for the arrow to reach its highest point, how much time passed after shooting the arrow until it reached the same height as before Troy shot it?

Problem 4.30 Dylan dives from a platform that is 40 feet high. Dylans mom wants to take a picture right when Dylan is at the same height as her. Dylan's height above the water after t seconds is $h(t) = -16t^2 + 8t + 40$. If Dylan's mom is sitting 10 feet above the water, how much time will pass before his mom takes the picture?

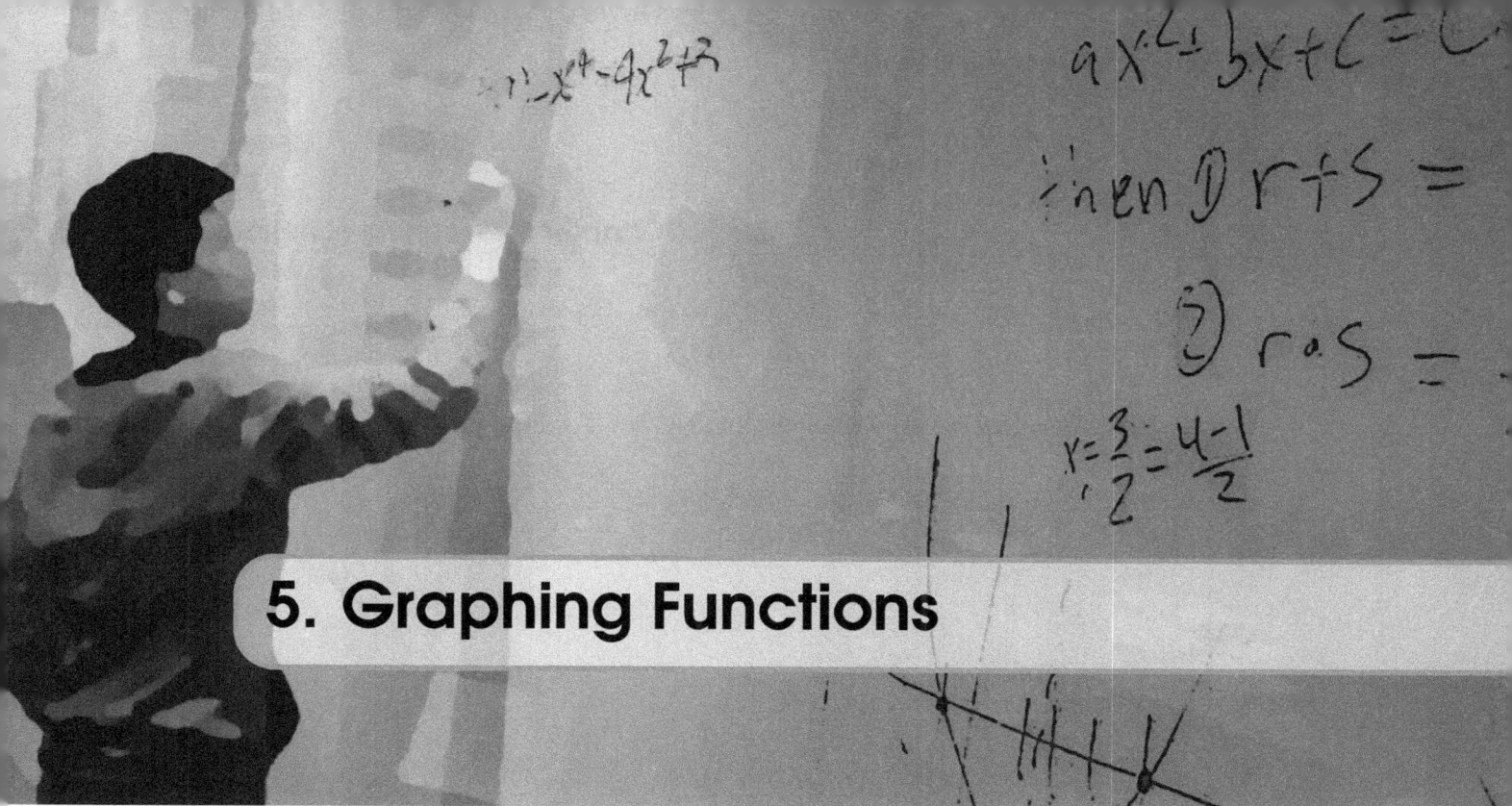

5. Graphing Functions

Graphing Equations and Translations

- Linear Equations:
    - $y = mx + b$ for a line with slope m and y-intercept b (slope-intercept form).
    - $y - k = m(x - h)$ for a line with slope m and point (h, k) (point-slope form).
- Quadratic Equations:
    - $y - k = a(x - h)^2$ for a quadratic equation with vertex (h, k).
- Translations:
    - Replacing x by $x - h$ in an equation shifts the graph h units right.
    - Replacing y by $y - k$ in an equation shifts the graph k units up.

Domain and Range

- Recall a function is a relation where each input has exactly on output.
- For a function, define:
    - the *domain* of the function as the types of input that are allowed. In particular, for graphs the domain is the possible x-values you can input.
    - the *range* of the function as the types of output that are achievable. In particular, for graphs, the range is the possible y-values that you can get as outputs for the function.

Rules of Exponents Revisited

- Exponents denote repeated multiplication. For example, $2^4 = 16$ because $2 \times 2 \times 2 \times 2 = 16$. Similarly $x^4 = x \times x \times x \times x$.

- Negative exponents denote repeated division. For example, $2^{-4} = \dfrac{1}{16}$ because $\dfrac{1}{2 \times 2 \times 2 \times 2} = \dfrac{1}{16}$.

- The fractional exponent $1/n$ denotes the nth root. For example, $16^{1/4} = 2$ because $\sqrt[4]{16} = 2$.

- Recall the following useful rules for exponents.
  - $x^0 = 1$ when $x \neq 0$.
  - $x^a \times x^b = x^{a+b}$.
  - $\dfrac{x^a}{x^b} = x^{a-b}$ or $\dfrac{x^a}{x^b} = \dfrac{1}{x^{b-a}}$.
  - $(x^a)^b = x^{ab}$.

- In particular these rules imply $x^{a/b} = \sqrt[b]{x^a}$.

5.1   Example Questions

Problem 5.1 Graph the following pairs on the same coordinate plane. Compare the two graphs.

(a) (i) $y = x$ and (ii) $y - 3 = -2(x - 2)$.

(b) (i) $y = x^2$ and (ii) $y + 2 = \dfrac{1}{2}(x - 1)^2$.

Problem 5.2 Graph the following equations by plotting points.

(a) $y = x^3$.

(b) $y = \sqrt{x}$

Problem 5.3 Graph the following equations.

(a) $y = 2x^3 - 6x^2 + 6x - 2$.

(b) $y = 2 - \sqrt{x-2}$.

Problem 5.4 What is the domain and range of the following?

(a) $y + 3 = 2(x+4)^2$.

(b) $y - 4 = -(x-2)^2/2$.

Problem 5.5 Find the domain and range of the following functions.

(a) $f(x) = (x-5)^5 + 5$.

(b) $f(x) = (x+2)^4 - 3$.

(c) $f(x) = 3\sqrt{x-4} + 5$.

Problem 5.6 Graph the following exponential functions by plotting points.

(a) $y = 2^x$.

(b) $y = 3^x$.

Problem 5.7 Exponential Function Graphing Continued

(a) Graph $y = \left(\dfrac{1}{2}\right)^x$. How does this compare to the graph of $y = 2^x$?

(b) Graph $y = \left(\dfrac{1}{3}\right)^x$. How does this compare to the graph of $y = 3^x$?

Problem 5.8 Exponents Revisited

(a) Calculate $\left(\dfrac{1}{3}\right)^{-3}$.

(b) Calculate $8^{-2/3}$.

(c) Calculate $9^{2/3}$ and express your answer in simplest radical form.

Problem 5.9 Graph the following and give the domain and range of each.

(a) $y = 2 + 2^{x-2}$

(b) $y = 3 \times \left(\dfrac{1}{3}\right)^x$

Problem 5.10 Suppose a certain type of bacterium divides once every hour. A certain petri dish starts with 4 bacteria. Write an equation representing the number of bacteria present after x hours and graph this equation.

5.2   Quick Response Questions

Problem 5.11 Which of the following gives the equation for the graph below?

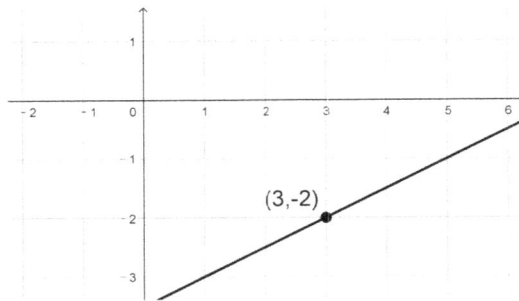

(A) $x = 3$

(B) $y - 2 = \frac{1}{2}(x - 3)$

(C) $y + 2 = \frac{1}{2}(x - 3)$

(D) $y + 2 = 2(x - 3)$

Problem 5.12 Which of the following gives the equation for the graph below?

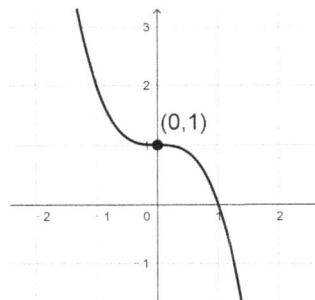

(A) $y = 1 + x^3$

(B) $y = 1 - x^3$

(C) $y = x^3 - 1$

(D) $y = -x^3$

Problem 5.13 Which of the following gives the equation for the graph below?

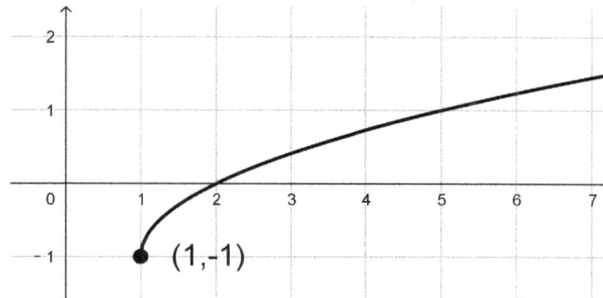

(A) $y = \sqrt{x-1} - 1$
(B) $y = (x-1)^2 - 1$
(C) $y = \sqrt{x-1} + 1$
(D) $y = (x-1)^2 + 1$

Problem 5.14 Which of the following gives the equation for the graph below?

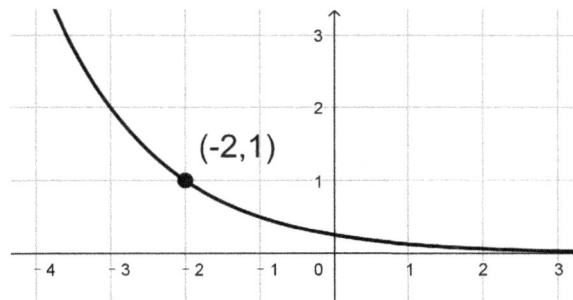

(A) $y = 2^x$
(B) $y = 0.5^x$
(C) $y = 2^{x+2}$
(D) $y = 0.5^{x+2}$

Problem 5.15 The range of the parabola $y+3=-2(x-4)^2$ is all y such that $y \leq L$ for some integer L. What is L?

Problem 5.16 What is the domain of $y = 2 - \sqrt{x-2}$?

(A)   All x such that $x \geq 2$
(B)   All y such that $y \geq 2$
(C)   All x such that $x \leq 2$
(D)   All x such that $y \leq 2$

Problem 5.17 What is the range of $y = 2 - \sqrt{x-2}$?

(A)   All x such that $x \geq 2$
(B)   All y such that $y \geq 2$
(C)   All x such that $x \leq 2$
(D)   All x such that $y \leq 2$

Problem 5.18 Which of the following is NOT equal to $x^{4/5}$?

(A) $\sqrt[5]{x^4}$
(B) $x^{-5/4}$
(C) $1/x^{-4/5}$
(D) $(x^{1/5})^4$

Problem 5.19 The graph $y = a^x$ (for a an integer) contains the points $(0,1)$ and $(3,64)$. What is a?

Problem 5.20 Is it true that ALL lines have a domain and range of all real numbers?

5.3 Practice Questions

Problem 5.21 Write the parabola $y = 2x^2 - 16x + 35$ in the form $y - k = a(x - k)^2$. How does the graph of the parabola compare to the graph of $y = x^2$? (You do not need to graph both parabolas, just explain how the graphs are related.)

Problem 5.22 Graph the equation $y = \dfrac{1}{x}$ by plotting points.

Problem 5.23 Graph the equation $y = \dfrac{1}{x-1} - 1$.

Problem 5.24 What is the domain and the range of the function $y = \dfrac{1}{x-1} - 1$?

Problem 5.25 Find the domain of the function $y = \sqrt{3 - x}$.

Problem 5.26 Graph the equations $y = 5^x$ and $y = \left(\dfrac{1}{5}\right)^x$ on the same coordinate plane.

Problem 5.27 Graph the equations $y = x^2$ and $y = 2^x$ on the same coordinate plane. For $x \geq 0$, which of the graphs seems to grow "faster"? You may want to consider large values of x!

Problem 5.28 Rewrite each of the following functions in the form $y = a^x$ for a rational number $a > 0$.

(a) $y = \dfrac{1}{2}2^x$.

(b) $y = 2^{-2x}$

Problem 5.29 Graph the equation $y = 2 - 2^x$.

Problem 5.30 Nickel-63 has a half-life of about 1 century. This means that after every century, half of the nickel-63 you have will decay. Suppose you start out with 2 kg of nickel-63. Write an equation that gives the amount (in kg) of nickel remaining after x centuries.

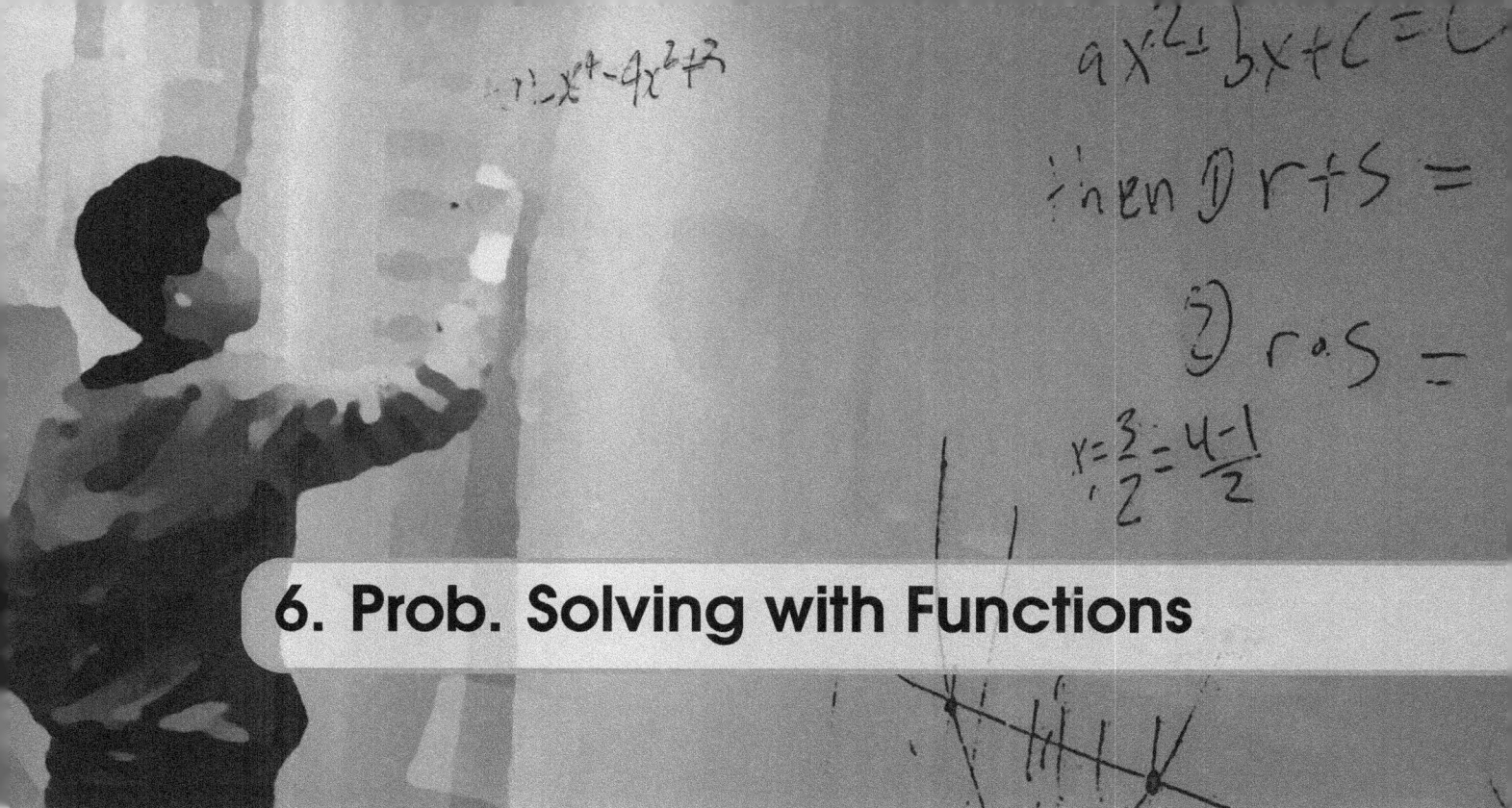

6. Prob. Solving with Functions

Review of Laws of Exponents

- $x^0 = 1$ when $x \neq 0$.
- $x^{-n} = \dfrac{1}{x^n}$ for integers $n \geq 0$.
- $x^{1/n} = \sqrt[n]{x}$ for integers $n \geq 1$.
- $x^a \times x^b = x^{a+b}$.
- $\dfrac{x^a}{x^b} = x^{a-b}$ or $\dfrac{x^a}{x^b} = \dfrac{1}{x^{b-a}}$.
- $\left(x^a\right)^b = x^{ab}$.

Review of Domain and Range

- Recall a function is a relation where each input has exactly on output.
- For a function, define:
    - the *domain* of the function as the types of input that are allowed. In particular, for graphs the domain is the possible x-values you can input.
    - the *range* of the function as the types of output that are achievable. In particular, for graphs, the range is the possible y-values that you can get as outputs for the function.

Modeling using Functions

- Given some data, we might want to come up with an equation that matches or approximates the data. This is called modeling.

- Depending on the scenario and on the data, different types of functions (linear, quadratic, exponential, or square root) can be used for the model.

6.1 Example Questions

Problem 6.1 Review of Exponents

(a) Calculate $4^{3/2}$.

(b) Calculate $\left(\dfrac{2}{3}\right)^{-3}$

(c) Simplify $(\sqrt[4]{5})^{10} \times \dfrac{1}{5}$ by writing it in the form 5^A for some number A.

Problem 6.2 Graph the equation $y = 4 - 3^{-x+2}$.

Problem 6.3 George starts with a stack of pennies. Each night before bed he makes a second stack with the same number of pennies as the first stack. Every morning me uses the two stacks to make one bigger stack of pennies. Suppose George starts with 3 pennies.

(a) If y denotes the number of pennies in George's stack, write an equation for y after x days.

(b) Equations of the form $y = C \times (1 + R)^x$ are examples of exponential growth. For your answer to part (a), explain the meaning of C and R.

(c) How many pennies are in George's stack after 7 days?

Problem 6.4 Sara bought a new car, worth $15,000$. After some research, she determined that for each year she owned the car, the car would be worth 20% less than the year before.

(a) If y denotes the value of Sara's car in dollars, write an equation for y, after Sara owns the car for x years.

(b) Equations of the form $y = C \times (1 - R)^x$ are examples of exponential decay. For your answer to part (a), explain the meaning of C and R.

(c) How much is Sara's car worth after 3 years?

Problem 6.5 Peter is given 1000 by his grandfather to invest. Peter finds a good investment opportunity that will give him 10% returns per year.

(a) If he gets his return once a year, he earns 10% of his investment at the end of the year. If he gets his return twice a year, he earns $10\% \div 2 = 5\%$ of his investment after 6 months and another 5% at the end of the year. Does this change the amount he earns in one year?

(b) Suppose that whenever he earns money, Peter can reinvest that money. If Peter gets returns once a year, write an equation for how much money Peter's investment is worth after x years.

(c) Suppose again that whenever he earns money, Peter can reinvest that money. If Peter gets returns twice a year, write an equation for how much money Peter's investment is worth after x years.

(d) Which scenario is best for Peter? Explain your answer.

Problem 6.6 Solve the following equations.

(a) $x^3 = 125$.

(b) $2^x = 8$.

(c) $9^x = 243$.

Problem 6.7 Solve the following equations.

(a) $\frac{1}{5} \times 5^x = 125$

(b) $3\left(\frac{1}{2}\right)^x = 192$

Problem 6.8 Consider the data graphed below:

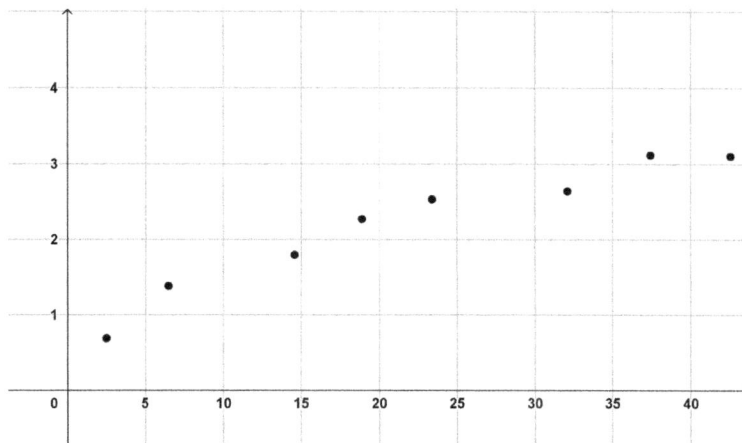

(a) If we want to model this data using one of the types of functions we have learned about (linear, quadratic, exponential, square root), which type of function do you think is most applicable? Why?

(b) Using your answer from part (a), come up with an equation modeling the data.

(c) Use your model to estimate the value of y when $x = 80$.

Problem 6.9 Consider the data in the following table

x	y
-2	0.11
-1	0.35
0	1.1
1	2.9
2	9.2

(a) If we want to model this data using one of the types of functions we have learned about (linear, quadratic, exponential, square root), which type of function do you think is most applicable? Why?

(b) Using your answer from part (a), come up with an equation modeling the data.

(c) Use your model to estimate the value of x when $y = 735$.

Problem 6.10 Kim and Harry just launched their new app. They were excited to get 400 downloads by day 2 and 1600 downloads by day 4.

(a) Kim thinks the downloads are growing exponentially. Help Kim find a model for their downloads!

(b) Harry thinks their downloads are growing quadratically, with a vertex when $x = 0$. Help Harry find a model for their downloads!

(c) Are either Kim or Harry wrong? Discuss the pros and cons of each model.

6.2 Quick Response Questions

Problem 6.11 In the equation $5 \times 2^x = 160$, what is x?

Problem 6.12 In the equation $18 \times 3^x = 2$, what is x?

Problem 6.13 What is y when $x = 4$, if $y = \dfrac{2}{5} \times 10^x$?

Problem 6.14 Which of the following is an exponential growth model?

(A) $y = 5 \times 1^x$

(B) $y = \dfrac{1}{2} \times 2^x$

(C) $y = 3 \times .99^x$

(D) None of the above

Problem 6.15 Drake had some pieces of paper. He split each in half. Then he split all of the resulting pieces in half one more time. If he had 48 pieces of paper in the end, with how many pieces did he start?

Problem 6.16 Donnie opened a savings account that gives her .3% monthly interest. Donnie's first deposit was of $300. Which of the following describes the amount of money in Donnie's account after x months?

(A) $y = 300 \times .3^x$

(B) $y = 300 \times .003^x$

(C) $y = 300 \times 1.3^x$

(D) $y = 300 \times 1.003^x$

Problem 6.17 In the exponential growth model $y = 2 \times 3^x$, what is R?

Problem 6.18 In the exponential decay model $y = 200 \times .7^x$, what is R?

Problem 6.19 Is the following an example of exponential decay?

$$y = 4 \times \left(\frac{8}{5}\right)^x$$

Problem 6.20 Lisa can type 50 words per minute at the beginning of her work day. She noticed that as the day goes on, she types less words per minute. Every hour her count of words per minute decreases approximately by 5%. How many words per minute can she type after 3 hours working? Round your answer to the nearest integer if necessary.

6.3 Practice Questions

Problem 6.21 Practice with Exponents

(a) Calculate $\left(\dfrac{16}{9}\right)^{-3/2}$.

(b) Write $\dfrac{\sqrt{2}}{8}$ in the form 2^K for some number K.

Problem 6.22 Graph the equation $y = \left(\dfrac{1}{2}\right)^{1-x}$.

Problem 6.23 Clark loves collecting comic books. Each year he increases the size of his collection by 50%. Clark's collection started with 64 comic books.

(a) Write an equation for the number of comic books in Clark's collection after x years.

(b) How many books are in Clark's collection after 5 years?

Problem 6.24 Lauri started the month with $100 in spending money. Each day she spent $1/3$ of her remaining money.

(a) Write an exponential decay model of the form $y = C \times (1-R)^x$ denoting how much money Lauri had left after x days. What is C and R?

(b) How many days was it until Lauri had less than $25 dollars remaining?

Problem 6.25 George invested $300 with a 8% interest rate compounded 4 times a year. Write an equation for the amount of money George has after x years?

Problem 6.26 Solve the equation $3^x = \dfrac{1}{81}$.

Problem 6.27 Solve $3^x + 3^x = 18$.

Problem 6.28 Troy wants to model some data using exponential growth, so he uses an equation of the form $y = C \times (1+R)^x$. The data contains the points $(0,5)$ and $(2,45)$. What is the equation for Troy's model?

Problem 6.29 Consider the data found below.

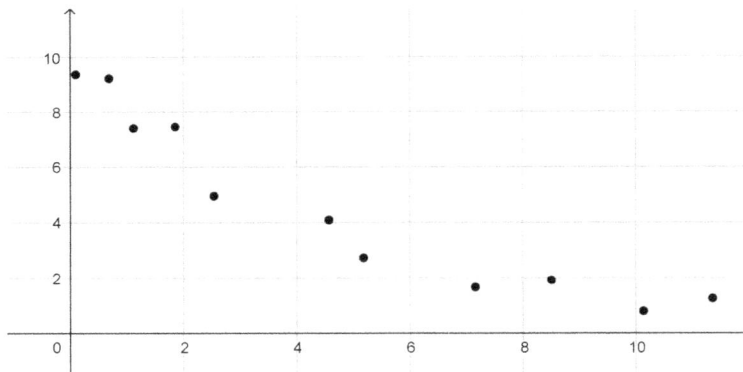

We want to build an exponential decay model for this data, using the equation $y = C \times (1-R)^x$.

(a) If C is an integer, estimate C.

(b) If R is of the form $L\%$ where L is a multiple of 10, estimate R.

Problem 6.30 Consider the following data again

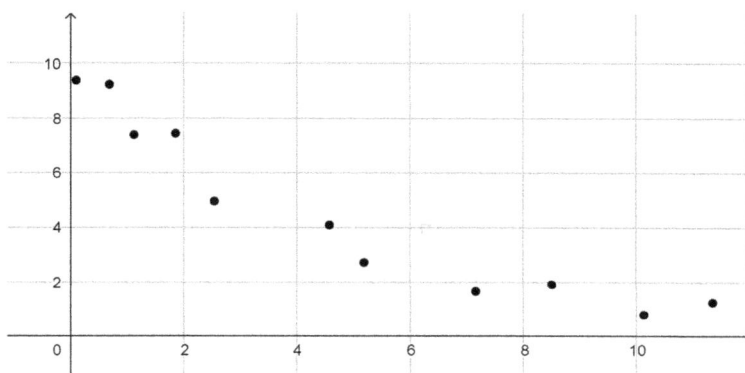

This time we want to build a quadratic model with vertex $(10, 1)$, so using an equation of the form $y - 1 = a(x - 10)^2$. If a is of the form $1/K$ for an integer K, give an equation for the model.

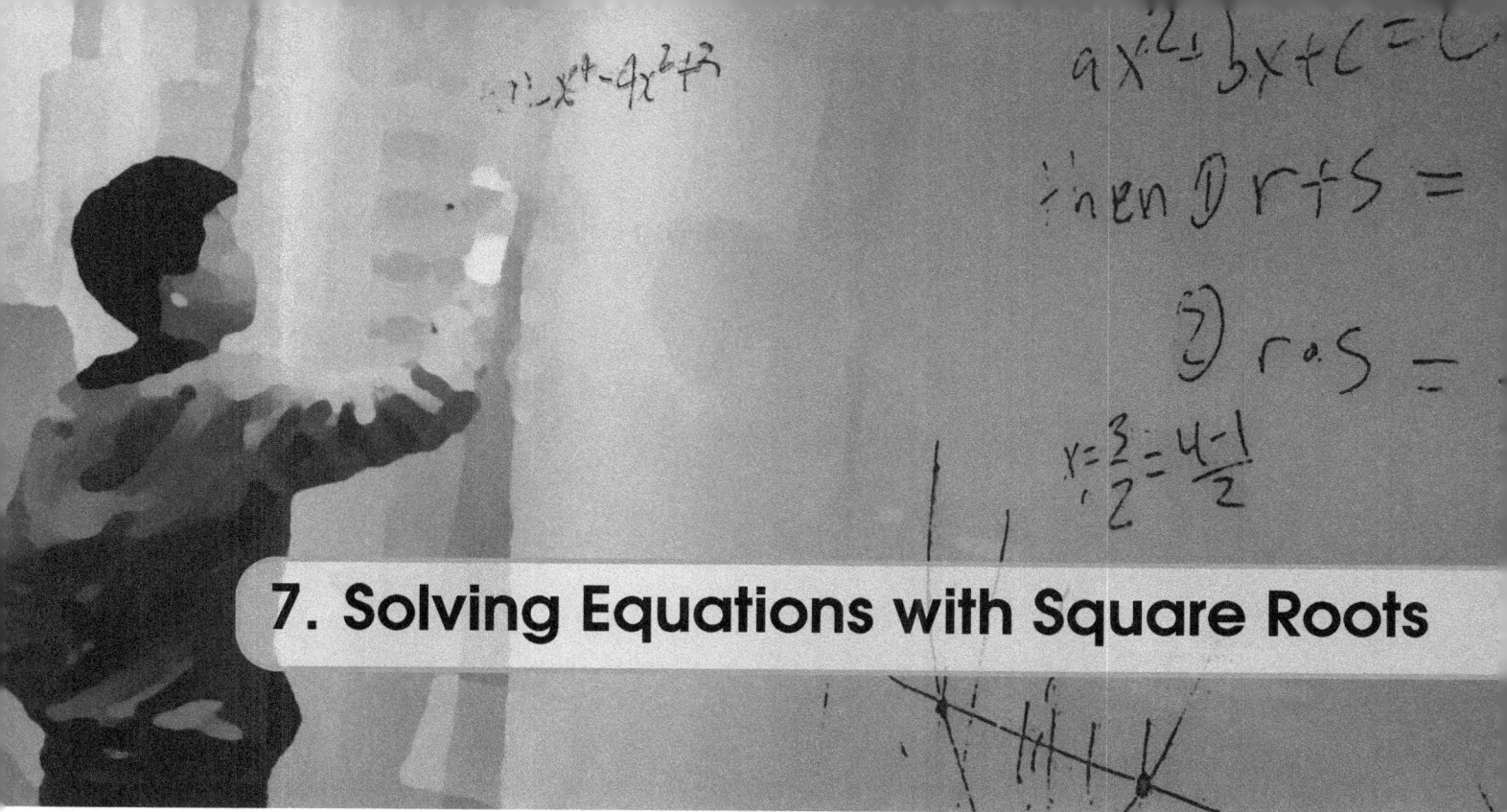

7. Solving Equations with Square Roots

7.1 Example Questions

Problem 7.1 Solving simple equations involving square roots

(a) Solve $x^2 = 9$.

(b) Solve $\sqrt{x} = 3$.

(c) Solve $\sqrt{x} = -3$.

Problem 7.2 Solve the following equations.

(a) $2\sqrt{x} = 6 - \sqrt{x}$.

(b) $\sqrt{x} + \sqrt{2x} = 1$.

Problem 7.3 Solve the following equations.

(a) $\sqrt{x-3} + 4 = 8$.

(b) $3 - \sqrt[3]{x+5} = 8$.

Problem 7.4 Solve the following by squaring both sides. Be careful about extraneous roots!

(a) $\sqrt{x^2 + 3x} = 2$

(b) Solve $\sqrt{2x^2 + 3x - 10} = x$.

Problem 7.5 Solve the equation $\sqrt{1 + \sqrt{2+x}} = 3$.

Problem 7.6 Solve the following equations by squaring twice.

(a) $\sqrt{x+3} - \sqrt{2-x} = 1$

(b) $\sqrt{2x} - \sqrt{2x - 10} = 4$.

Problem 7.7 Recall that $x^{a/b} = (\sqrt[b]{x})^a$. Solve the following equations for x.

(a) $x^{-1/3} = \dfrac{1}{3}$

(b) $(r-1)^{-2/3} = \dfrac{1}{4}$.

Problem 7.8 Solve the equation $\sqrt{|x|+x} = 4$.

Problem 7.9 Jenna and Olivia both try to model the growth of their basil plant. Jenna models the growth using the equation $y = x+1$ while Olivia models using the equation $y = \sqrt{x^2 - 2x + 3}$.

(a) Do their models ever agree?

(b) For what x is Jenna's predicted growth larger than Olivia's model?

Problem 7.10 Find the domain and range of $y = \sqrt{x^2 - 4x + 8}$.

7.2 Quick Response Questions

Problem 7.11 In the equation $\sqrt{x} = 6$, what is x?

Problem 7.12 How many real solutions does the equation $\sqrt{x+1} = -4$ have?

Problem 7.13 Which of the following expressions is equivalent to $\left(\sqrt[4]{a}\right)^3$?

(A) $a^{\frac{4}{3}}$

(B) $a^{-\frac{4}{3}}$

(C) $a^{\frac{3}{4}}$

(D) $a^{-\frac{3}{4}}$

Problem 7.14 Which of the following expressions is equivalent to $\left(\frac{8}{125}\right)^{-\frac{2}{3}}$?

(A) $\left(\frac{8}{125}\right)^{\frac{3}{2}}$

(B) $\frac{25}{4}$

(C) $\left(\sqrt[2]{\frac{8}{125}}\right)^3$

(D) $-\frac{4}{25}$

Problem 7.15 Is it true that for all values of x we have $\sqrt{x^2} = x$?

Problem 7.16 What is the domain of the function $y = \sqrt{x^2 + 4x + 4}$?

(A) $x \leq 2$
(B) $x \geq 2$
(C) All real numbers
(D) None of the above

Problem 7.17 In the equation $\sqrt{x + 2} = 4$, what is x?

Problem 7.18 In the equation $x^{\frac{2}{3}} = 16$, what is x?

Problem 7.19 In the equation $\sqrt[3]{x + 1} = 2$, what is x?

Problem 7.20 In the equation $4\sqrt{x} = 3 + \sqrt{x}$, what is x?

7.3 Practice Questions

Problem 7.21 Solve the equation $\sqrt{x-3} = 1 + \sqrt{3}$.

Problem 7.22 Solve $\sqrt{3x} - \sqrt{9x} = -6$.

Problem 7.23 Solve the equation $\sqrt[3]{2x+4} = \sqrt[3]{x+10}$.

Problem 7.24 Solve the following equations.

(a) $\sqrt{x^2 + 4x - 3} = x$

(b) $\sqrt{3x^2 - 8} = x$

Problem 7.25 Solve the equation $\sqrt[3]{10 - \sqrt{x+2}} = 2$.

Problem 7.26 Solve the equation $\sqrt{2x} - \sqrt{2x - 10} = 2$.

Problem 7.27 Solve the equation $(5a - 1)^{-4/5} = 1$.

Problem 7.28 Solve the equation $\sqrt{x-4} + |x^2 - 16| = 0$. Remember the properties of square roots and absolute values!

Problem 7.29 Jenna and Olivia are back at it building models. They are having an argument about their newest models for bacteria growth. Jenna claims the growth is $y = \sqrt{2 \times 4^x}$ while Olivia claims it is $y = 2^{x+0.5}$. Where do their models agree?

Problem 7.30 Find the domain and range of $y = \sqrt{8x - x^2}$.

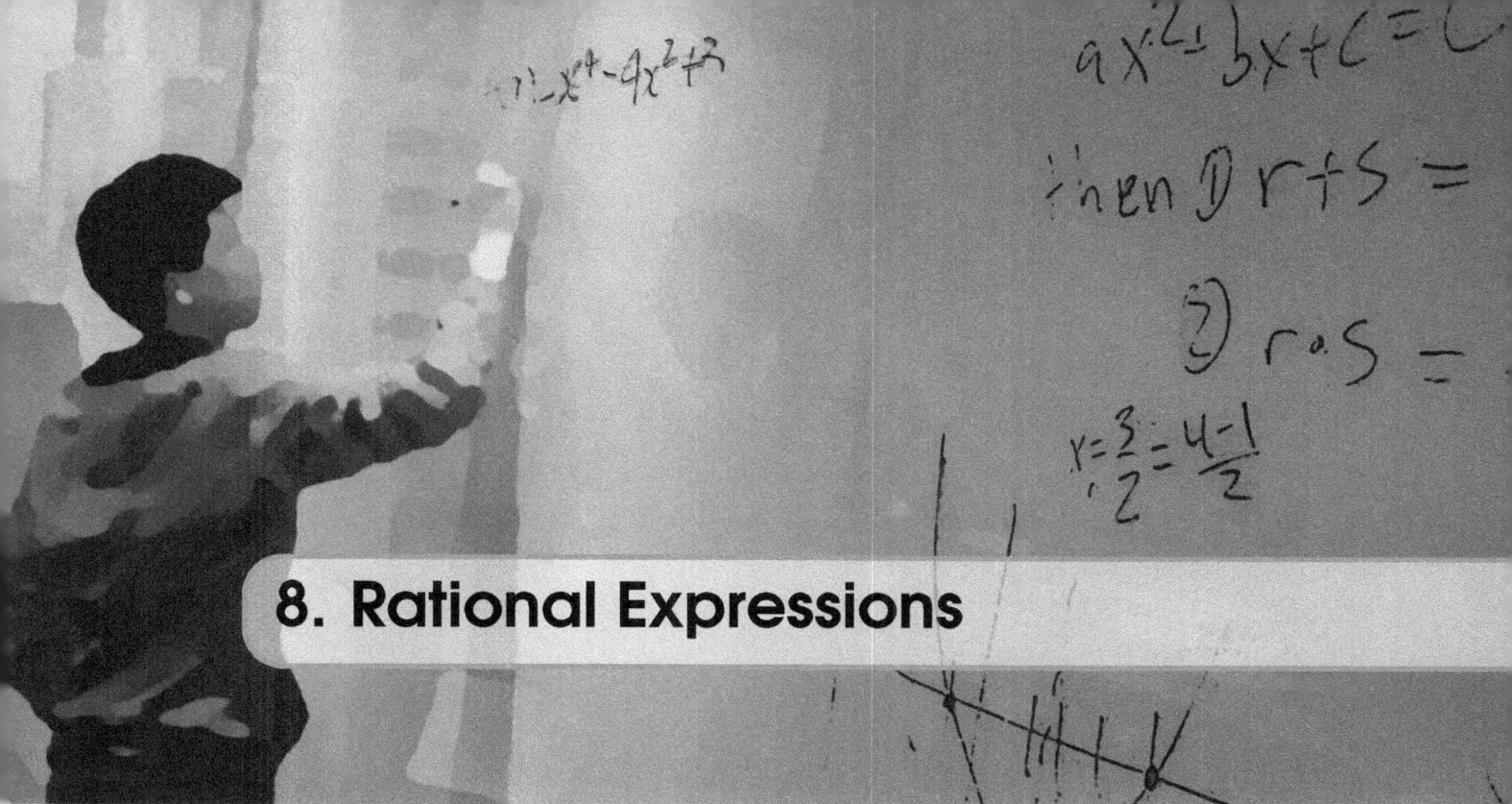

8. Rational Expressions

Review of Long Division

- Long division for numbers was a systematic way to perform division.
- When dividing a number (called the dividend) by another number (called the divisor) we can calculate the quotient and the remainder.
- For example, $11 \div 3$ has a quotient of 3 and a remainder of 2. Note we can write this in multiple ways:
    - $\dfrac{11}{3} = 3 + \dfrac{2}{3}$, with the form quotient plus remainder over dividend.
    - $11 = 3 \times 3 + 2$, with the form quotient times dividend plus remainder.

Long Division for Polynomials

- When dividing polynomials, we can use polynomial long division to calculate quotients and remainders.
- If we are dividing a polynomial $P(x)$ by another polynomial $D(x)$ we can write:
    - $P(x) = D(x)Q(x) + R(x)$, where $\deg R(x) < \deg D(x)$. Here $Q(x)$ is the quotient, and $R(x)$ is the remainder.
    - or $\dfrac{P(x)}{D(x)} = Q(x) + \dfrac{R(x)}{Q(x)}$.

Asymptotes

- When graphing equations, sometimes the graph will get closer and closer to a

---

line, but never actually touch it. The line it gets closer and closer to is called an asymptote.

- Asymptotes are classified as either
    - horizontal,
    - vertical, or
    - slant or oblique. (Slant asymptotes are straight lines.)
- For example, the graph of $y = \dfrac{1}{x}$ has asymptotes of $x = 0$ (vertical) and $y = 0$ (horizontal).

8.1 Example Questions

Problem 8.1 Expand the following.

(a) $(t-2)(t^2 - t + 3)$

(b) $(x^3 + 2x + 3)(x^2 + 4x + 2)$

Problem 8.2 Review of Long Division, Quotients, and Remainders

(a) Calculate $789 \div 6$ using long division.

(b) Write $789 = Q \times 6 + R$ where Q is the quotient and R is the remainder when dividing 789 by 6.

(c) Write $122 = Q \times 4 + R$ where Q is the quotient and R is the remainder when dividing 122 by 4.

Problem 8.3 Division with Polynomials

(a) Divide $(a-4)$ by a.

(b) Verify that $\dfrac{x^2}{x-2} = (x+2) + \dfrac{4}{x-2}$.

Problem 8.4 Calculate the quotient and remainder of $(x^3+2) \div (x+1)$.

Problem 8.5 Calculate $(2v^4 + v^3 - 6v^2 - v + 1) \div (2v+1)$.

Problem 8.6 Calculate $(x^3 + 2x^2 + x + 1) \div (x^2 + x + 1)$.

Problem 8.7 What is the domain of the following functions?

(a) $y = \dfrac{x^2 - 2x - 3}{x-1}$

(b) $y = \dfrac{1}{x^2 + x + 1}$

Problem 8.8 Graph the equation $y = \dfrac{x^2 + 2x - 3}{x-1}$.

Problem 8.9 Graph the equation $y = \dfrac{x^2 + 2x + 4}{x^2 - 4}$. What are the asymptotes?

Problem 8.10 Graph the equation $y = \dfrac{x^2}{x - 2}$. What are the asymptotes?

8.2 Quick Response Questions

Problem 8.11 What is the quotient when 354 is divided by 11?

Problem 8.12 What is the remainder when 354 is divided by 11?

Problem 8.13 Which of the following is NOT an answer to the question: What is $123 \div 7$?

(A)   17 remainder 4
(B)   rounded to the nearest integer is 17
(C) $17\frac{4}{7}$
(D) ≈ 17.57

Problem 8.14 Which of the equations corresponds to the statement, $3x \div (x-2)$ has quotient 3 and remainder 6?

(A) $3x(x-2) = 3x^2 - 6$
(B) $3x = 6(x-2) + 3$
(C) $3x = 3(x-2) + 6$
(D) $x - 2 = 3(3x) + 6$

Problem 8.15 If we divide one polynomial by another, does the remainder always have to be a constant (a number)?

Problem 8.16 Is it true that the remainder of $(x^5 + 4x^4 + 3x + 2) \div (x^2 + 3x + 5)$ is $3x^2 + 5x + 3$?

Problem 8.17 The remainder when $x^2 - 3x + 5$ is divided by $x + 2$ is an integer r. What is r?

Problem 8.18 The domain of the function $y = \dfrac{2x^2 + 7x + 6}{2x + 3}$ is all x except $x = L$ for a number L. What is L? Round your answer to the nearest tenth if necessary.

Problem 8.19 Consider the x-intercept or x-intercepts of the function $y = \dfrac{2x^2 + 7x + 6}{2x + 3}$. What is the sum of all of the x-intercepts? (If there are none the sum is 0, if there is only one x-intercept the sum is just that intercept.)

Problem 8.20 Which of the following is true for the graph of $y = x + 2 + \frac{2}{(x-1)(x+3)}$?

    (A)   There is one vertical asymptote and one slant asymptote.
    (B)   There are two horizontal asymptotes and one slant asymptote. $+3$
    (C)   There are two vertical asymptotes and one horizontal asymptote.
    (D)   There are two vertical asymptotes and one slant asymptote.

8.3 Practice Questions

Problem 8.21 Expand $(3x^2 + 4x + 2)(x^2 - 8)$.

Problem 8.22 Consider the expression $x^2 - x + 1 + \dfrac{-2}{x+1}$.

(a) Write the expression as one fraction with denominator of $x + 1$.

(b) What is the quotient and remainder when $x^3 - 1$ is divided by $x + 1$?

Problem 8.23 Calculate $\dfrac{t^2 - 5t - 36}{t + 4}$. Can you do this without using polynomial long division?

Problem 8.24 What is $(x^5 - 1) \div (x - 1)$?

Problem 8.25 Calculate $\dfrac{x^3 + 3x + 2}{x + 2}$.

Problem 8.26 Calculate $\dfrac{x^3 + 3x + 2}{x^2 - x}$.

Problem 8.27 What is the domain and range of the function $y = \dfrac{x^2 - 5x - 36}{x + 4}$?

Problem 8.28 Graph the equation $y = \dfrac{x^4 - 2x^2 + 1}{x^2 - 1}$.

Problem 8.29 Graph the equation $y = \dfrac{2 + x}{2 - x}$. What are the asymptotes?

Problem 8.30 What are the asymptotes of the graph of $y = \dfrac{x^3 + 3x + 2}{x^2 - x}$?

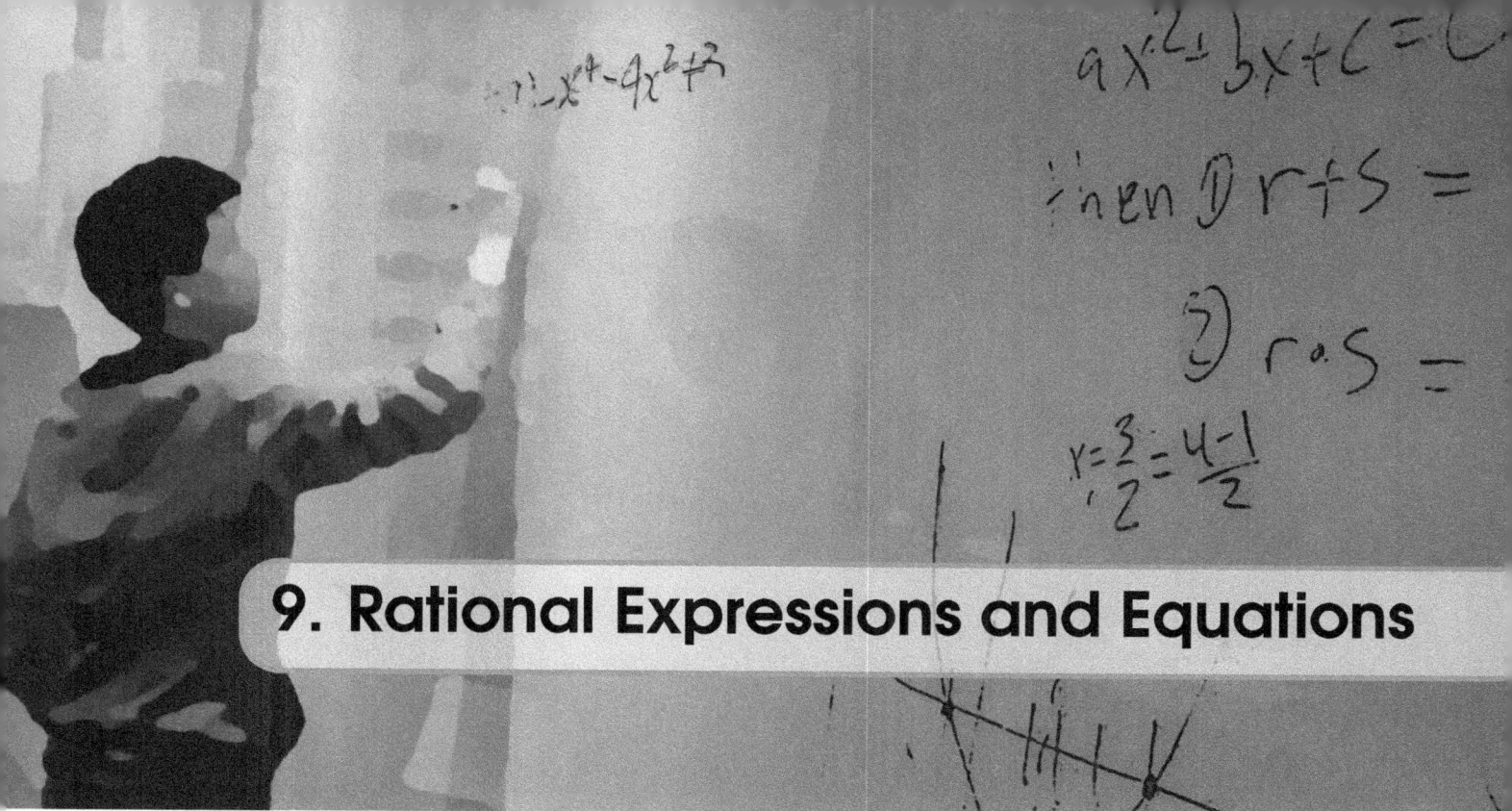

9. Rational Expressions and Equations

Review of Long Division for Polynomials

- When dividing polynomials, we can use polynomial long division to calculate quotients and remainders.
- If we are dividing a polynomial $P(x)$ by another polynomial $D(x)$ we can write:
  - $P(x) = D(x)Q(x) + R(x)$, where $\deg R(x) < \deg D(x)$. Here $Q(x)$ is the quotient, and $R(x)$ is the remainder.
  - or $\dfrac{P(x)}{D(x)} = Q(x) + \dfrac{R(x)}{Q(x)}$.

Theorems about Long Division for Polynomials

- **Polynomial Remainder Theorem**: Suppose you are dividing a polynomial $P(x)$ by a linear divisor $(x - a)$. Then
  - The remainder of $P(x) \div (x - a)$ is equal to $P(a)$ (recall $P(a)$ is the polynomial evaluated when $x = a$).
- **Factor Theorem**: Suppose you are solving a polynomial equation $P(x) = 0$ where $P(x)$ is a polynomial. Then the following two statements are equivalent:
  - $x = r$ is a solution to the equation $P(x) = 0$.
  - $(x - r)$ is a factor of the polynomial $P(x)$.
- Note the Factor Theorem can be useful for factoring cubics, as if you know one factor you can then reduce the problem to factoring a quadratic.

Rational Root Theorem

- Suppose you are solving a polynomial equation $P(x) = 0$ where $P(x)$ has degree n and $P(x) = A \times x^n + \cdots + B$. ($P(x)$ has leading coefficient A and constant term B.) Then:
    - If $x = r = \dfrac{P}{Q}$ is a rational solution to the equation $P(x) = 0$ then P is a factor of $\pm B$ and Q is a factor or $\pm A$.
- For example, the only possible rational roots of $2x^4 - x^3 + 7x^2 + 5x + 3 = 0$ are $x = \pm 1, \pm 3, \pm \dfrac{1}{2}, \pm \dfrac{3}{2}$.

9.1   Example Questions

Problem 9.1 What is the quotient and remainder when $x^5 + 3x^2 - 2x + 1$ is divided by $x^2 + x + 1$?

Problem 9.2 Verify the Polynomial Remainder Theorem for the following. That is show that the remainder of $P(x) \div (x - a)$ is equal to $P(a)$.

(a) $P(x) = x^3 + x^2 + x + 6$ and $a = -2$.

(b) $P(x) = x^3 + 2x^2 + x - 3$ and $a = 5$.

Problem 9.3 Solve the following equations using the given hint.

(a) $x^3 - 5x^2 + 5x + 3 = 0$ if $x = 3$ is one solution.

(b) $x^4 + x^3 - 6x^2 - 4x + 8 = 0$ if $x = 1$ and $x = -2$ are solutions.

Problem 9.4 Suppose the remainder when $(kx^2 - x + 2) \div (x - 5)$ is -53. What is k?

Problem 9.5 Consider $P(x) = 2x^3 - ax^2 - bx - 1$. Suppose the remainder when $P(x) \div (x + 1)$ is -10 and the remainder when $P(x) \div (x + 2) = -37$. What are a and b?

Problem 9.6 Practice with the Rational Root Theorem

(a) What are the possible rational roots of $2x^3 - 11x^2 + 4x + 5 = 0$?

(b) Use the possible rational roots to help solve equation $2x^3 - 11x^2 + 4x + 5 = 0$.

Problem 9.7 Write each of the following rational functions in the form $y = \dfrac{P(x)}{Q(x)}$ for polynomials $P(x)$ and $Q(x)$, simplifying where possible. What is the domain?

(a) $y = \dfrac{1}{x^2 + x} + \dfrac{x^2}{x + 1}$.

(b) $y = \dfrac{x}{x + 1} - \dfrac{2}{x + 2}$.

Problem 9.8 Solve the equation $\dfrac{1}{x} + \dfrac{1}{x + 1} + \dfrac{1}{x - 1} = 0$.

Problem 9.9 Solve the equation $\dfrac{2}{x^2+3x+2} = 1 - \dfrac{x}{x^2-1}$.

Problem 9.10 Suppose you expand $(x-1)^7$ to get a polynomial (with degree 7). What is the sum of all the coefficients of this polynomial?

9.2 Quick Response Questions

Problem 9.11 According to the Rational Root Theorem, is $x = \frac{3}{2}$ a possible root for $6x^3 + 2x^2 + 3x + 8 = 0$?

Problem 9.12 According to the Rational Root Theorem, is $x = \frac{22}{9}$ a possible root for $18x^5 + 24x^2 - 82x^2 + 9x + 44 = 0$?

Problem 9.13 Is $(x + 1)$ a factor of $x^3 + x^2 - x + 1$?

Problem 9.14 Is $(x - 2)$ a factor of $x^3 - 5x^2 + 8x - 4$?

Problem 9.15 Use the Remainder Theorem to find the remainder of

$$(x^4 + 2x^3 - 4x + 2) \div (x + 1)$$

without doing long division.

Problem 9.16 Use the Remainder Theorem to find the remainder of

$$(x^5 + 3x^2 + 5) \div (x - 2)$$

without doing long division.

Problem 9.17 If $x = -1$ is a solution to $x^3 + 4x^2 + 4x + 1 = 0$, how many real solutions does the equation have?

Problem 9.18 Solve for x in $\dfrac{1}{x} = \dfrac{2}{x} + 2$.

Problem 9.19 Solve for x in $3 - \dfrac{1}{2x} = \dfrac{1}{x} + 1$

Problem 9.20 Solve for x in $\dfrac{3}{x+1} = \dfrac{2}{x+1} + 5$

9.3 Practice Questions

Problem 9.21 Find the quotient and remainder when $(9x^3 - 6x^2 + 2x + 8) \div (3x - 1)$.

Problem 9.22 Find the remainder when $(x^4 + 2x^3 - 2x^2 - 9x - 6) \div [(x+1)(x-2)]$.

Problem 9.23 Solve the equation $x^3 - 2x^2 - 9x + 18 = 0$ if you are given that $x = \pm 3$ are both roots.

Problem 9.24 Suppose $(x + 3)$ is a factor of $x^2 - x + k$. What is k?

Problem 9.25 Consider $P(x) = x^3 + ax^2 + bx + 1$. Suppose $P(1) = 3$ and $P(2) = 17$. What is the remainder when $P(x)$ is divided by $(x + 2)$?

Problem 9.26 According to the Rational Root Theorem, what are the possible rational roots of $4x^4 + 5x^3 + x^2 - 6$?

Problem 9.27 Write the expression $\dfrac{x+1}{x^2+3} - \dfrac{1}{x}$ as one single fraction.

Problem 9.28 Solve the equation $\dfrac{2}{x-2} + \dfrac{1}{x+1} - \dfrac{1}{x-1} = 1$.

Problem 9.29 Solve the equation

$$\frac{x}{x^2 - 5x + 6} = \frac{x - 2}{x^2 - x - 6}.$$

Problem 9.30 Expand $(3x - 1)^6$ to get $Ax^6 + Bx^5 + Cx^4 + Dx^3 + Ex^2 + Fx + G$. What is $A + B + C + D + E + F$?

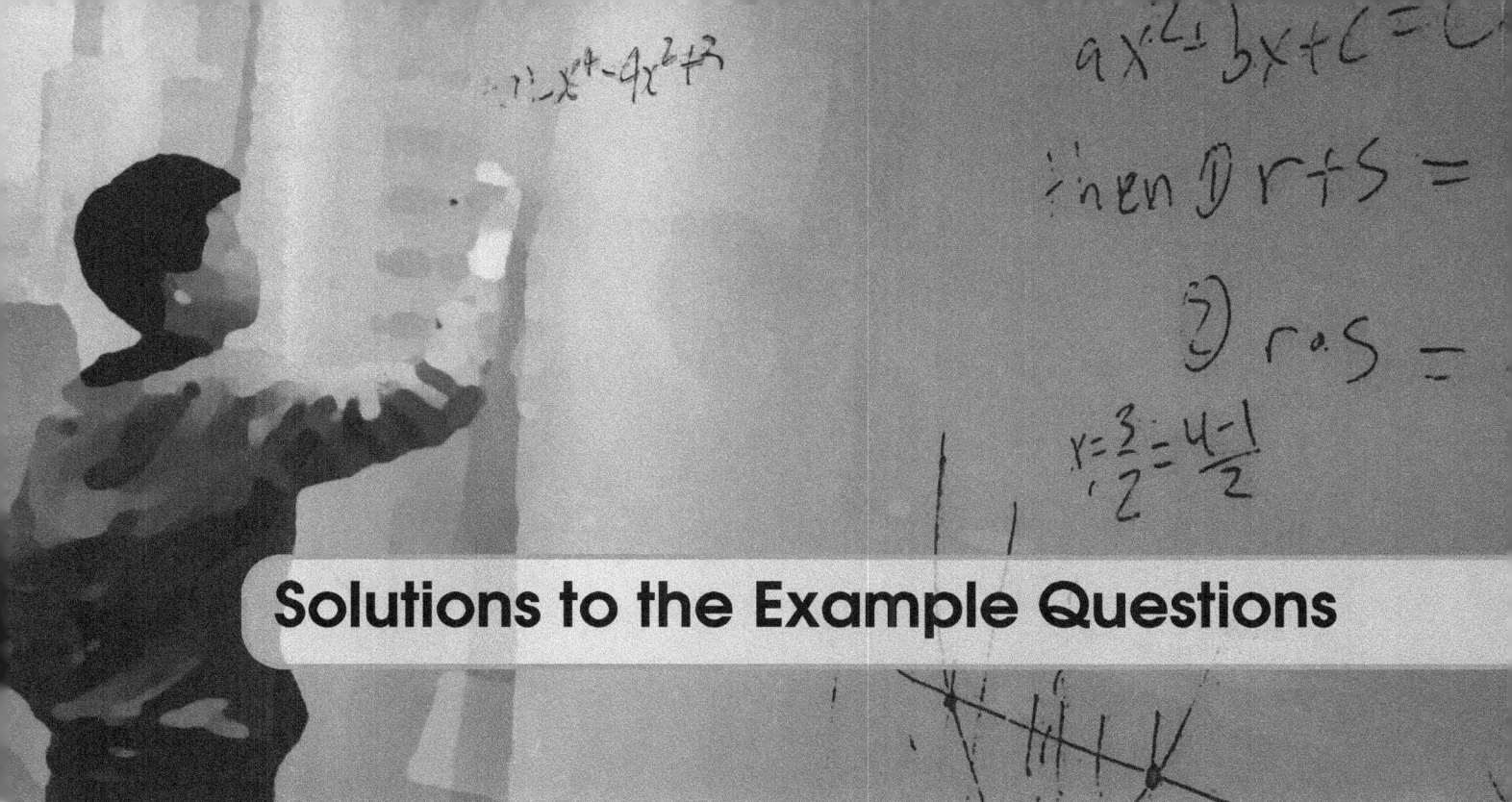

Solutions to the Example Questions

In the sections below you will find solutions to all of the Example Questions contained in this book.

Quick Response and Practice questions are meant to be used for homework, so their answers and solutions are not included. Teachers or math coaches may contact Areteem at info@areteem.org for answer keys and options for purchasing a Teachers' Edition of the course.

1   Solutions to Chapter 1 Examples

Problem 1.1 Review of Distributive Property

(a) Expand $2x(x^2 + 3)$.

| Answer |

$2x^3 + 6x$

| Solution |

Distributing the $2x$ we get $2x \times x^2 + 2x \times 3 = 2x^3 + 6x$.

(b) Expand $-3xy(x^3 + 2xy + y^2)$.

| Answer |

$-3x^4y - 6x^2y^2 - 3xy^3$

| Solution |

Distributing the $-3xy$ we have $-3xy \times x^3 + (-3xy) \times 2xy + (-3xy) \times y^2 = -3x^4y - 6x^2y^2 - 3xy^3$.

Problem 1.2 Factor the following by pulling out common terms.

(a) $2x^3 - 4x$.

| Answer |

$2x(x^2 - 2)$

| Solution |

Note that $2x^3$ and $4x$ have a $2x$ in common. Pulling this out we have $2x(x^2 - 2)$.

(b) $12x^4 + 3x^3 + 6x^2$.

| Answer |

$3x^2(4x^2 + x + 2)$

Solution

Examining the terms we see they have $3x^2$ as a common factor. Therefore we have $12x^4 + 3x^3 + 6x^2 = 3x^2(4x^2 + x + 2)$.

Problem 1.3 Review of expanding binomials (FOIL)

(a) Expand $(x+3)(x+4)$.

Answer

$x^2 + 7x + 12$.

Solution

Using FOIL we have $x \times x + x \times 4 + 3 \times x + 3 \times 4 = x^2 + 4x + 3x + 12 = x^2 + 7x + 12$.

(b) Expand $(2x+3)(x-1)$.

Answer

$2x^2 + x - 3$

Solution

Using FOIL we have $2x \times x + 2x \times (-1) + 3 \times x + 3 \times (-1) = 2x^2 - 2x + 3x - 3 = 2x^2 + x - 3$.

(c) Expand $(3x^2 + y)(y + 2)$.

Answer

$3x^2y + 6x^2 + y^2 + 2y$

Solution

Using FOIL we have $3x^2 \times y + 3x^2 \times 2 + y \times y + y \times 2 = 3x^2y + 6x^2 + y^2 + 2y$.

Problem 1.4 Factor the following.

(a) $x^2 + 5x + 6$.

Answer

$(x+2)(x+3)$

Solution

Using Reverse FOIL we want two numbers with a product of 6 and sum of 5, which are $2, 3$. Hence $x^2 + 5x + 6 = (x+2)(x+3)$.

(b) $x^2 - x - 12$.

Answer

$(x-4)(x+3)$

Solution

Using Reverse FOIL we want two numbers with a product of -12 and sum of -1. These numbers are -4 and 3, giving a factoring of $x^2 - x - 12 = (x-4)(x+3)$.

(c) $2x^2 - 7x + 3$.

Answer

$(2x-1)(x-3)$

Solution

Using Reverse FOIL we want two numbers with product 2 and two numbers with product 3, such that multiplying and adding gives -7. We see the numbers must be $2, 1$ and $-3, -1$ with $2 \times -3 + 1 \times -1 = -7$. This leads to the factoring $2x^2 - 7x + 3 = (2x-1)(x-3)$.

Problem 1.5 Solve the following equations

(a) $x^2 + 2x = 0$.

Answer

$x = 0$ and $x = -2$

Solution

Pulling out the common term of x we have $x^2 + 2x = x(x + 2) = 0$ so either $x = 0$ or $x + 2 = 0$. Hence the solutions are $x = 0$ and $x = -2$.

(b) $x^2 - 3x + 2 = 0$.

Answer

$x = 1$ and $x = 2$

Solution

Factoring using Reverse FOIL we have $x^2 - 3x + 2 = (x - 1)(x - 2) = 0$ so $x - 1 = 0$ or $x - 2 = 0$. Therefore $x = 1$ and $x = 2$ are the solutions.

(c) $2x^2 - 15x - 27 = 0$.

Answer

$x = 9$ and $x = -1.5$

Solution

Factoring gives $(2x - 3)(x - 9) = 0$ so $2x - 3 = 0$ or $x - 9 = 0$. This gives the two solutions of $x = 9$ and $x = -1.5$.

Problem 1.6 Albert was playing baseball and hit a ball straight up with an initial speed of 47 feet per second. The height of the ball was modeled by the equation $h = -16t^2 + 47t + 3$. How long did it take for the ball to come back down and hit the ground?

Answer

3 seconds

Solution

We know the height is given by $h = -16t^2 + 47t + 3$. Since the height on the ground is $h = 0$, we want to solve $-16t^2 + 47t + 3 = 0$ or $16t^2 - 47t - 3 = 0$ after multiplying both sides by -1. Factoring we have $(16t + 1)(t - 3) = 0$ so either $16t + 1 = 0$ or $t - 3 = 0$.

Hence either $t = -1/16$ or $t = 3$. As the ball cannot hit the ground before it is thrown, we throw out $t = -1/16$. Therefore it took 3 seconds for the ball to come back down.

Problem 1.7 Solve the following.

(a) $y(y-6) + y(y+1) + 25 = 2y(y+5) - 5$.

Answer

$y = 2$

Solution

Distributing gives $y^2 - 6y + y^2 + y + 25 = 2y^2 + 10y - 5$ so $2y^2 - 5y + 25 = 2y^2 + 10y - 5$. Collecting all like terms we have (noticing that the $2y^2$ cancels from both sides) $-15y = -30$ so $y = 2$.

(b) $2y(y+3) = 8$.

Answer

$y = 1$ or $y = -4$.

Solution

Distributing and collecting all the terms on the left-hand side we have $2y^2 + 6y - 8 = 0$. After dividing by 2 and factoring we have $y^2 + 3y - 4 = (y+4)(y-1) = 0$ so $y = 1$ or $y = -4$.

Problem 1.8 Practice with Polynomials

(a) Calculate $2x(x^2 + 3x + 2) - 3x^2(x+2)$.

Answer

$-x^3 + 4x$

Solution

Distributing we have $2x^3 + 6x^2 + 4x - 3x^3 - 6x^2$. Combining like terms gives $-x^3 + 4x$.

(b) Expand $(x^2 + 2)(x^2 - 3x + 2)$.

Answer

$x^4 - 3x^3 + 4x^2 - 6x + 4$

Solution

We have the above is equal to $x^4 - 3x^3 + 2x^2 + 2x^2 - 6x + 4$ so combining like terms we have $x^4 - 3x^3 + 4x^2 - 6x + 4$.

Problem 1.9 Solve the following polynomial equations.

(a) $-2x^3 - 2x^2 + 12x = 0$

Answer

$x = 0, x = 2, x = -3$

Solution

First note we can divide by -2 to simplify the equation to $x^3 + x^2 - 6x = 0$. Factoring we have $x(x^2 + x - 6) = x(x + 3)(x - 2) = 0$ so $x = 0$, $x + 3 = 0$, or $x - 2 = 0$. Gives gives solutions of $x = 0, -3, 2$.

(b) $2x^2(x - 1) + 9x(x - 1) - 5x + 5 = 0$.

Answer

$x = 1, x = 0.5$, or $x = -5$.

Solution

Note the first two expressions share a common factor of $x - 1$. Rewriting the left-hand side we have

$$2x^2(x - 1) + 9x(x - 1) - 5(x - 1) = (x - 1)[2x^2 + 9x - 5].$$

Hence after factoring we have $(x - 1)(2x - 1)(x + 5) = 0$ so $x = 1$, $x = 0.5$, or $x = -5$.

Problem 1.10 Hank is planning building a jacuzzi in his backyard. He initially planned to make the jacuzzi 4 feet by 6 feet. After thinking some more, he decided he wanted to double the area of his jacuzzi. If he still wants the jacuzzi to be 2 feet longer than it is wide, how many feet should he add to the dimensions of the jacuzzi to double the area?

Answer

2

Solution

The original planned dimensions of Hank's jacuzzi are 4 feet by 6 feet with area 24 square feet. Hence he wants the jacuzzi with area $24 \times 2 = 48$ square feet. Let x be the feet Hank should add to the length and width, so the new jacuzzi has dimensions $4 + x$ feet by $6 + x$ feet. Thus we want

$$(4+x)(6+x) = 48 \text{ or } x^2 + 10x + 24 = 48 \text{ or } x^2 + 10x - 24 = 0.$$

Factoring gives $x^2 + 10x - 24 = (x + 12)(x - 2) = 0$ so $x = -12$ or $x = 2$. Since the answer of -12 feet does not make sense, Hank should add 2 feet to the dimensions of his jacuzzi to double the area.

2   Solutions to Chapter 2 Examples

Problem 2.1 Finding our first pattern. Expand the following:

(a) $(a+b)^2$

Answer

$a^2 + 2ab + b^2$

Solution

Expanding we have

$$\begin{aligned}
(a+b)^2 &= (a+b)(a+b) \\
&= a^2 + ab + ba + b^2 \\
&= a^2 + 2ab + b^2.
\end{aligned}$$

(b) $(a+b)^3$

Answer

$a^3 + 3a^2b + 3ab^2 + b^3$

Solution

We know from part a) that $(a+b)^2 = a^2 + 2ab + b^2$. Hence expanding we get

$$\begin{aligned}
(a+b)^3 &= (a+b)(a+b)^2 \\
&= (a+b)(a^2 + 2ab + b^2) \\
&= a^3 + 2a^2b + ab^2 + a^2b + 2ab^2 + b^3 \\
&= a^3 + 3a^2b + 3ab^2 + b^3.
\end{aligned}$$

(c) $(a+b)^4$

Answer

$a^4 + 4a^3b + 6a^2b^2 + 4ab^3 + b^4$

Solution

We know from part b) that $(a+b)^3 = a^3 + 3a^2b + 3ab^2 + b^3$. Hence expanding we get

$$
\begin{aligned}
(a+b)^4 &= (a+b)(a+b)^3 \\
&= (a+b)(a^3 + 3a^2b + 3ab^2 + b^3) \\
&= a^4 + 3a^3b + 3a^2b^2 + ab^3 + a^3b + 3a^2b^2 + 3ab^3 + b^4 \\
&= a^4 + 4a^3b + 6a^2b^2 + 4ab^3 + b^4.
\end{aligned}
$$

Problem 2.2 Review of Factoring.

(a) Factor $x^2 - 3x - 4$.

Answer

$(x-4)(x+1)$

Solution

Using Reverse Foil we have $x^2 - 3x - 4 = (x-4)(x+1)$.

(b) Factor $2x^2 + 8x + 8$.

Answer

$2(x+2)^2$

Solution

First we have a common factor of 2: $2x^2 + 8x + 8 = 2(x^2 + 4x + 4)$. Then using Reverse FOIL or recognizing a perfect square we have $2(x^2 + 4x + 4) = 2(x+2)^2$.

(c) Factor $x^2 - 9$

Answer

$(x-3)(x+3)$

Solution

Using either Reverse FOIL or recognizing a difference of two squares we have $x^2 - 9 = (x-3)(x+3)$.

Problem 2.3 Solve the following equations.

(a) $x^2 - x + \dfrac{1}{4} = 0$.

Answer

$x = \dfrac{1}{2}$

Solution

We can factor the left-hand side as a perfect square. Hence $x^2 - x + \dfrac{1}{4} = \left(x - \dfrac{1}{2}\right)^2 = 0$ so $x - \dfrac{1}{2} = 0$ and hence $x = \dfrac{1}{2}$.

(b) Solve $x^4 - 16 = 0$.

Answer

$x = \pm 2$

Solution

Using the difference of two squares we have $x^4 - 16 = (x^2)^2 - 4^2 = (x^2 + 4)(x^2 - 4)$. We can further factor the $x^2 - 4$ to $(x+2)(x-2)$. Hence we have $(x^2 + 4)(x+2)(x-2) = 0$. As $x^2 + 4$ cannot be zero, we either have $x + 2 = 0$ so $x = -2$ or $x - 2 = 0$ so $x = 2$.

Problem 2.4 Solve the following equations involving exponents.

(a) $x^2 = 25$.

Answer

$x = \pm 5$

Solution

$x^2 = 25$ so either $x = 5$ or $x = -5$.

(b) $x^3 = -27$.

Answer

$x = -3$

Solution

$x^3 = -27$ so we must have $x = -3$.

(c) $x^4 = -64$.

Answer

No solutions.

Solution

$x^4 = -64$, but x^4 is always positive. Hence there are no solutions.

Problem 2.5 Expand the following.

(a) $(x+3)(x^2 - 3x + 9)$

Answer

$x^3 + 27$

Solution

We have $(x+3)(x^2 - 3x + 9) = x^3 - 3x^2 + 9x + 3x^2 - 9x + 27 = x^3 + 27$.

(b) $(a+b)(a^2 - ab + b^2)$

Answer

$a^3 + b^3$

Solution

Expanding gives $(a+b)(a^2-ab+b^2) = a^3 - a^2b + ab^2 + a^2b - ab^2 + b^3 = a^3 + b^3$.

(c) $(a-b)(a^2+ab+b^2)$

Answer

$a^3 - b^3$

Solution

Expanding gives $(a-b)(a^2+ab+b^2) = a^3 + a^2b + ab^2 - a^2b - ab^2 - b^3 = a^3 - b^3$.

Problem 2.6 Factor Theorem Practice

(a) Suppose $(x-2)(x+k) = 0$ has solutions of 2 and 3. What is k?

Answer

-3

Solution

Since the equation has solutions of $x = 2$ and $x = 3$ we know that $(x-2)$ and $(x-3)$ are factors. Hence matching up the factors we see $(x-3)$ must be the same as $(x+k)$ so $k = -3$.

(b) Is $(x-2)$ a factor of $x^4 - 2x^3 + 2x^2 - 8x + 8$?

Answer

Yes.

Solution

If $(x-2)$ is a factor, then substituting $x = 2$ should give 0. Checking, $2^4 - 2 \times 2^3 + 2 \times 2^2 - 8 \times 2 + 8 = 0$, so $(x-2)$ is indeed a factor.

Problem 2.7 Practice the Factoring Patterns!

(a) Factor $8x^3 + 125$.

Answer

$(2x+5)(4x^2 - 10x + 25)$

Solution

We recognize this as $(2x)^3 + 5^3$ so using the sum of two cubes formula we have $(2x + 5)((2x)^2 - (2x)(5) + 5^2) = (2x+5)(4x^2 - 10x + 25)$.

(b) $x^8 - 1$

Answer

$(x^4 + 1)(x^2 + 1)(x+1)(x-1)$

Solution

Note $x^8 = (x^4)^2$ and $x^4 = (x^2)^2$ so repeatedly using the difference of two squares we have $x^8 - 1 = (x^4 + 1)(x^4 - 1) = (x^4 + 1)(x^2 + 1)(x^2 - 1) = (x^4 + 1)(x^2 + 1)(x+1)(x-1)$.

(c) $8x^3 + 12x^2 + 6x + 1$

Answer

$(2x+1)^3$

Solution

We have $8x^3 = (2x)^3$ and $1 = 1^3$, so we suspect that the above is $(2x+1)^3$. Double checking we have $(2x+1)^3 = (2x)^3 + 3 \times (2x)^2 + 3 \times 2x + 1 = 8x^3 + 12x^2 + 6x + 1$ as needed.

Problem 2.8 Solve the following equations.

(a) $x^4 - 6x^2 + 9 = 0$.

Answer

$x = \pm\sqrt{3}$

Solution

Factoring we have $x^4 - 6x^2 + 9 = (x^2 - 3)^2 = 0$ so $x^2 - 3 = 0$ so $x^2 = 3$. Hence $x = \pm\sqrt{3}$.

(b) $(x + 2)^2 = 8$.

Answer

$x = -2 \pm 2\sqrt{2}$

Solution

First note $(x + 2)^2 = 8$ tells us that $x + 2 = \pm\sqrt{8} = \pm 2\sqrt{2}$. Hence $x = -2 \pm 2\sqrt{2}$.

Problem 2.9 Factoring Tricks

(a) Factor $x^2 - 1 + x^3 - 1$.

Answer

$(x - 1)(x^2 + 2x + 2)$

Solution

Grouping as $(x^2 - 1)(+ (x^3 - 1)$ we have $(x + 1)(x - 1) + (x - 1)(x^2 + x + 1)$ using the difference of squares and cubes. As each has a factor of $(x - 1)$ we get $(x - 1)[(x + 1) + (x^2 + x + 1)] = (x - 1)(x^2 + 2x + 2)$.

(b) Factor $x^3 + x^2 + x + 1$.

Answer

$(x + 1)(x^2 + 1)$

Solution

Grouping as $(x^3 + x^2) + (x + 1)$ we have $x^2(x + 1) + (x + 1) = (x + 1)(x^2 + 1)$.

Problem 2.10 Solving Equations Tricks

(a) Solve the equation $x^2 - 6x + 9 = 5$.

Answer

$x = 3 \pm \sqrt{5}$

Solution

Recognize the left-hand side as the perfect square $(x-3)^2$. Hence we need $(x-3)^2 = 5$ so $x - 3 = \pm\sqrt{5}$ and hence $x = 3 \pm \sqrt{5}$.

(b) Solve the equation $xy + x - 2y - 2 = 0$.

Answer

$x = 2$ or $y = -1$

Solution

Using either Reverse FOIL or grouping, note we have $xy + x - 2y - 2 = (x-2)(y+1) = 0$. Hence $x - 2 = 0$ o $x = 2$ or $y = -1$. Note this means as if $x = 2$ (and y is anything) the equation is true, and similarly if $y = -1$ (and x is anything) the equation is true.

3   Solutions to Chapter 3 Examples

Problem 3.1 Review of Solving Quadratic Equations

(a) $x^2 + 6x + 8 = 0$

Answer

$x = -2, x = -4$

Solution

Factoring we have $x^2 + 6x + 8 = (x+2)(x+4) = 0$ so $x+2 = 0$ so $x = -2$ or $x+4 = 0$ so $x = -4$.

(b) $x^2 + 6x + 9 = 3$

Answer

$x = -3 \pm \sqrt{3}$

Solution

We recognize the left-hand side as $(x+3)^2$, so we have $(x+3)^2 = 3$ so $x+3 = \pm\sqrt{3}$ and therefore $x = -3 \pm \sqrt{3}$.

Problem 3.2 Solve the following equations by "completing the square".

(a) $x^2 - 4x - 5 = 0$

Answer

$x = -1, x = 5$

Solution

To complete the square we need to add $(-4/2)^2 = (-2)^2$. Hence we have

$$x^2 - 4x - 5 = x^2 - 4x + (-2)^2 - (-2)^2 - 5 = (x^2 - 4x + 4) - 9 = (x-2)^2 - 9.$$

Thus $(x-2)^2 - 9 = 0$ so $(x-2)^2 = 9$. Thus $x - 2 = \pm 3$ so $x = 2 \pm 3$ which gives $x = 5$ or $x = -1$.

(b) $x^2 - 4x - 1 = 0$

Answer

$x = 2 \pm \sqrt{5}$

Solution

To complete the square we add $(-4/2)^2 = 4$. Thus $x^2 - 4x + 4 - 4 - 1 = 0$ so $(x-2)^2 - 5 = 0$. Therefore $(x-2) = \pm\sqrt{5}$ and hence $x = 2 \pm \sqrt{5}$.

Problem 3.3 Solve the equation $2x^2 + 9x + 3 = 0$ by completing the square.

Answer

$x = \dfrac{-9 \pm \sqrt{57}}{4}$

Solution

Here we must first divide by 2, giving $x^2 + \dfrac{9}{2}x + \dfrac{3}{2} = 0$. Therefore we must add $\left(\dfrac{9}{4}\right)^2 = \dfrac{81}{16}$ to complete the square. This gives

$$x^2 + \frac{9}{2}x + \frac{81}{16} - \frac{81}{16} + \frac{3}{2} = 0 \Rightarrow \left(x + \frac{9}{4}\right)^2 = \frac{57}{16}.$$

Therefore $x + \dfrac{9}{4} = \pm\dfrac{\sqrt{57}}{4}$ and hence $x = \dfrac{-9 \pm \sqrt{57}}{4}$.

Problem 3.4 Solve the equation $ax^2 + bx + c = 0$ by completing the square.

Answer

$x = \dfrac{-b \pm \sqrt{b^2 - 4ac}}{2a}$

Solution

Dividing by a we have $x^2 + \dfrac{b}{a}x + \dfrac{c}{a} = 0$. To complete the square we add $\left(\dfrac{b}{2a}\right)^2 = \dfrac{b^2}{4a^2}$:

$$x^2 + \frac{b}{a}x + +\frac{b^2}{4a^2} - \frac{b^2}{4a^2} + \frac{c}{a} = 0 \Rightarrow \left(x + \frac{b}{2a}\right)^2 = \frac{b^2 - 4ac}{4a^2}.$$

Therefore,

$$\left(x+\frac{b}{2a}\right)=\pm\frac{\sqrt{b^2-4ac}}{2a}\Rightarrow x=\frac{-b\pm\sqrt{b^2-4ac}}{2a}.$$

Note this gives us the quadratic formula.

Problem 3.5 Find solutions to the following using the quadratic formula.

(a) $3x^2+4x-4=0$

Answer

$x=-2, x=\dfrac{2}{3}$

Solution

We have $a=3$, $b=4$, and $c=-4$ so using the quadratic formula we have

$$x=\frac{-4\pm\sqrt{4^2-4\times3\times(-4)}}{2\times3}=\frac{-4\pm\sqrt{64}}{6}=\frac{-4\pm8}{6}.$$

Hence $x=\dfrac{-4+8}{6}=\dfrac{2}{3}$ or $\dfrac{-4-8}{6}=-2$.

(b) $x^2-6x+2=0$

Answer

$x=3\pm\sqrt{7}$

Solution

We have $a=1$, $b=-6$, and $c=2$ so using the quadratic formula we have

$$x=\frac{-(-6)\pm\sqrt{(-6)^2-4\times1\times2}}{2}=\frac{6\pm\sqrt{28}}{2}=\frac{6\pm2\sqrt{7}}{2}.$$

Hence $x=3\pm\sqrt{7}$.

Problem 3.6 Graph the equation $y=x^2+6x+5$. Label the vertex, x-intercepts, and y-intercept.

Answer

x-intercepts: $-5,-1$, y-intercept: 5, vertex: $(-3,-4)$

Solution

We first see the y-intercept is 5. Factoring we have $x^2 + 6x + 5 = (x+5)(x+1)$ so the x-intercepts are -5 and -1. Completing the square gives $y = x^2 + 6x + 3^2 - 3^2 + 5 = (x+3)^2 - 4$ so the vertex is $(-3, -4)$. This gives the graph

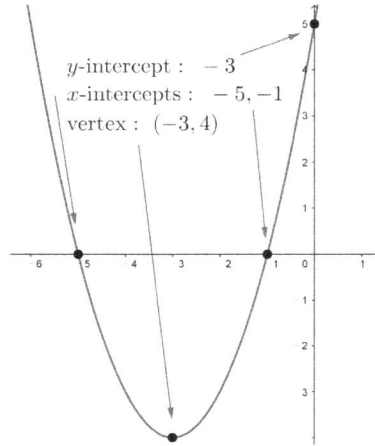

Problem 3.7 Graph the equation $y = a(x-h)^2 + k$

Answer

Answers may vary

Solution

The parabola opens upwards when $a > 0$ and downwards when $a < 0$. It has a vertex of (h, k). If $|a|$ is larger, the graph is narrower, while if $|a|$ is smaller the graph is wider.

Problem 3.8 Graph the equation $y = -2x^2 + x - 5$. Label the vertex and any intercepts. Is the vertex a maximum or minimum?

Answer

x-intercepts: None, y-intercept: -5, vertex: $(1/4, -39/8)$

Solution

The y-intercept is -5. Completing the square we have

$$-2x^2 + x - 5 = -2\left(x^2 - \frac{1}{2}x + \frac{5}{2}\right) = -2\left(x^2 - \frac{1}{2}x + \frac{1}{16} - \frac{1}{16} + \frac{5}{2}\right) = -2\left(x - \frac{1}{4}\right)^2 - \frac{39}{8}.$$

Hence the parabola opens downwards with vertex $(1/4, -39/8)$ so there are no x-intercepts. This gives the graph shown below:

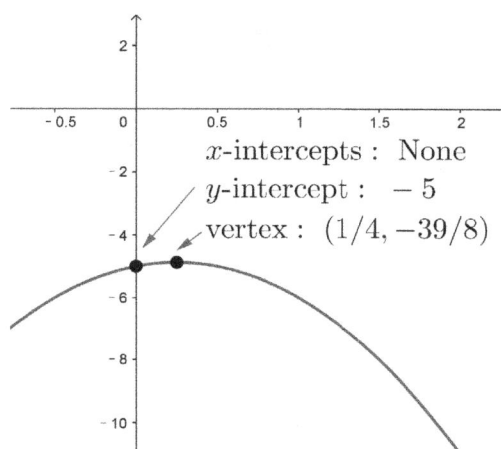

x-intercepts : None

y-intercept : -5

vertex : $(1/4, -39/8)$

Problem 3.9 For the equation $y = 4x - x^2$ find the intercepts and vertex. Is the vertex a maximum or minimum?

Answer

x-intercepts: $0, 4$, y-intercept: 0, vertex: $(2, 4)$ a maximum

Solution

First note the y-intercept is 0 and factoring $x(4 - x)$ we have x-intercepts of 0 and 4. Completing the square we get

$$-(x^2 - 4x) = -(x^2 - 4x + 4 - 4) = -(x - 2)^2 + 4.$$

Therefore the parabola has vertex $(2, 4)$. Since it opens downwards, the vertex is a maximum.

Problem 3.10 During their whale watching tour, Jason took a video of a dolphin jumping out of the water. He found out he could model the height of the dolphin during its jump using the quadratic equation $j(t) = 32t - 16t^2$. What was the maximum height of the dolphin during the jump?

Answer

16 feet

Solution

Completing the square we have $32t - 16t^2 = -16(t^2 - 2t) = -16(t^2 - 2t + 1 - 1) = -16(t-1)^2 + 16$. Therefore the parabola opens downwards, with a vertex of $(1, 16)$. Hence the maximum height is 16 feet.

4 Solutions to Chapter 4 Examples

Problem 4.1 Practice and Review using the Quadratic Formula

(a) Solve the equation $x^2 + 4x - 21 = 0$.

Answer

$3, -7$

Solution

Of course we can use the quadratic formula to find the solutions. However, remember that factoring can often be faster! We have $x^2 + 4x - 21 = (x+7)(x-3) = 0$ so $x = -7$ and $x = 3$ are the solutions.

(b) Solve the equation $2x^2 - 2x - 5 = 0$.

Answer

$\dfrac{1 \pm \sqrt{11}}{2}$

Solution

This equation does not factor. Using the quadratic formula (with $a = 2, b = -2, b = -5$) we have

$$x = \frac{-(-2) \pm \sqrt{(-2)^2 - 4 \times 2 \times (-5)}}{2 \times 2} = \frac{2 \pm \sqrt{44}}{4} = \frac{2 \pm 2\sqrt{11}}{4} = \frac{1 \pm \sqrt{11}}{2}$$

as our solutions.

Problem 4.2 Graph the equation $y = x^2 - 2x$. Be sure to label the vertex and any intercepts.

Answer

x-intercepts: $0, 2$, y-intercept: 0, vertex: $(1, -1)$

Solution

We see the y-intercept is 0. Factoring we have $x^2 - 2x = x(x-2)$ so the x-intercepts are

0 and 2. Completing the square we have $y = x^2 - 2x + 1 - 1$ so $y = (x-1)^2 - 1$ and hence the vertex is $(1, -1)$. This gives the graph below:

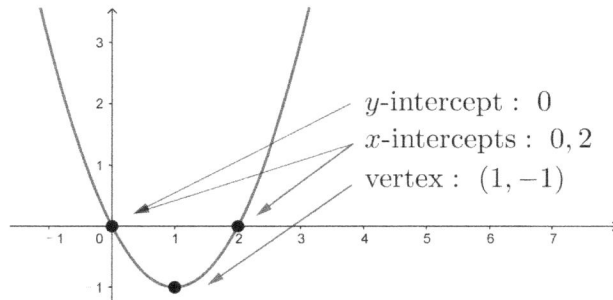

Problem 4.3 Consider the graph of the equaton $y = x^2 - 2x$.

(a) Do you notice any symmetry in the graph? How does this symmetry relate to the x-intercepts and the vertex?

Answer

The graph is the same left and right of the vertex.

Solution

Note the vertex is right between the two x-intercepts. We say the parabola has a $x = 1$ as a line of symmetry.

(b) More formally, we would say that $x = 1$ is a line of symmetry for the graph $y = x^2 - 2x$. This means that the y value for $x = 1 + k$ is the same as the y value for $x = 1 - k$ for any real number k. Verify this statement.

Solution

If we plug in $x = 1 + k$ we get

$$y = (1+k)^2 - 2(1+k) = k^2 + 2k + 1 - 2 - 2k = k^2 + 1.$$

If we plug in $x = 1 - k$ we get

$$y = (1-k)^2 - 2(1-k) = k^2 - 2k + 1 - 2 + 2k = k^2 + 1.$$

Note in both cases we get the same result as needed.

Problem 4.4 Consider general graphs of parabolas (so equations of the form $y = ax^2 + bx + c$).

(a) Do all parabolas have two x-intercepts? If not, what are the possibilities for the number of x-intercepts?

Answer

No: 2, 1, and 0 x-intercepts are possible

(b) The parabola $y = ax^2 + bx + c$ has "x-intercepts" of $\dfrac{-b \pm \sqrt{b^2 - 4ac}}{2a}$. What part of this equation leads to the case where there are 1 or 0 x-intercepts?

Answer

$b^2 - 4ac$, called the discriminant

Solution

Note we can only take the square root of a non-negative numbers. Hence if $b^2 - 4ac < 0$ there are no x-intercepts.

Similarly, if $b^2 - 4ac = 0$, we have $\dfrac{-b \pm 0}{2a} = \dfrac{-b}{2a}$, so there is only one x-intercept.

Problem 4.5 How many solutions do each of the following equations have? Use the discriminant.

(a) $x^2 + 3x + 2 = 0$.

Answer

2

Solution

The discriminant is $3^2 - 4 \times 1 \times 2 = 9 - 8 = 1$. Since this is positive, there are 2 solutions.

(b) $7x^2 + 14x + 8 = 0$.

Answer

0

Solution

The discriminant is $14^2 - 4 \times 7 \times 8 = 196 - 224 = -28$. Since this is negative, there are 0 solutions.

(c) $18x^2 - 21x - 7 = 0$.

Answer

2

Solution

The discriminant is $(-21)^2 - 4 \times 18 \times (-7) = 441 + 4 \times 18 \times 7$. Note without calculating further we know this is positive, so there are 2 solutions.

Problem 4.6 Dylan wants to throw a ball straight up in the air with a height (in feet) given by $h = -16t^2 + 35t + 5$. His friend Bob is filming horizontally out of his apartment's balcony at a height of 20 feet. How many times will the ball appear on the camera?

Answer

Twice

Solution

The ball will appear on the camera when the height is 20. Therefore we want $20 = -16t^2 + 35t + 5$, or $-16t^2 + 35t - 15 = 0$. Note the question just asks how many times the ball appears on the camera (not the actual times) so we can use the discriminant. We have the discriminant equals

$$35^2 - 4 \times (-16) \times (-15) = 1225 - 960 = 268,$$

so we know there are 2 solutions. Thus the ball appears on camera twice.

Problem 4.7 Carrie has an 8 by 10 inch photo she wants to have framed. The photo should fit evenly in the frame (same amount of space on each side), and she wants the area of the frame to be 150% the area of the photo. How big of a frame should Carrie buy?

Answer

10 inches by 12 inches

Solution

The photo has an area of $8 \times 10 = 80$ square inches.

Problem 4.8 Tony just got a new house, and had to spend most of his savings on a down payment. After he bought the house, he also had to spend more of his savings for moving, new furniture, etc. He modeled the balance of his savings account using the equation $S = \frac{1}{2}w^2 - 10x + 200$ where S is the amount in his account and w is the number of weeks after he bought he house. Does Tony run out of money?

Answer

No.

Solution

For Tony to run out of money we must have $0 = \frac{1}{2}w^2 - 10x + 200$. Note the discriminant is

$$(-10)^2 - 4 \times \frac{1}{2} \times 200 = 100 - 400 = -300 < 0.$$

Therefore the quadratic equation has no solutions, meaning that Tony never runs out of money.

Problem 4.9 A rectangle has a perimeter of 20 inches. What is the maximum area that this rectangle could have?

Answer

25 square inches

Solution

Let x be the length of the rectangle. Since the perimeter is 20, the length plus width is 10, so the width of the rectangle is $10 - x$. If we let y denote the area, then we have the equation $y = x(10 - x)$. Hence the graph of the area is a parabola. This parabola has x-intercepts of 0 and 10, hence the vertex occurs when $x = (0 + 10) \div 2 = 5$. Hence the maximum area is $5(10 - 5) = 25$, which occurs when the length and the width are both 5.

Problem 4.10 Consider the equation $3x^2 + bx + 12 = 0$. For what values of b does this equation have 2 real solutions?

Answer

$b > 12$ or $b < -12$

Solution

For two solutions the discriminant must be positive. Hence $b^2 - 4 \times 3 \times 12 > 0$ or $b^2 > 144$. Therefore we need $b > 12$ or $b < -12$.

5   Solutions to Chapter 5 Examples

Problem 5.1 Graph the following pairs on the same coordinate plane. Compare the two graphs.

(a) (i) $y = x$ and (ii) $y - 3 = -2(x - 2)$.

Solution

The graphs are shown below.

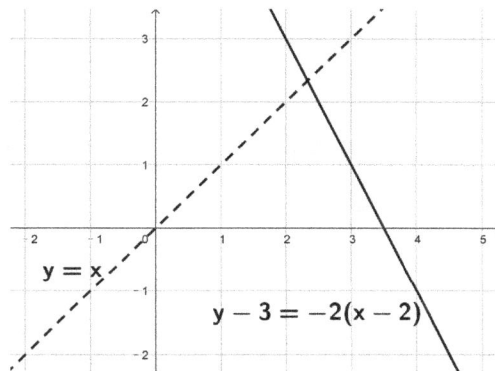

Note the second graph is shifted horizontally $+2$ and vertically $+3$ as well as flipped upside down and made narrower.

(b) (i) $y = x^2$ and (ii) $y + 2 = \dfrac{1}{2}(x - 1)^2$.

Solution

The graphs are shown below.

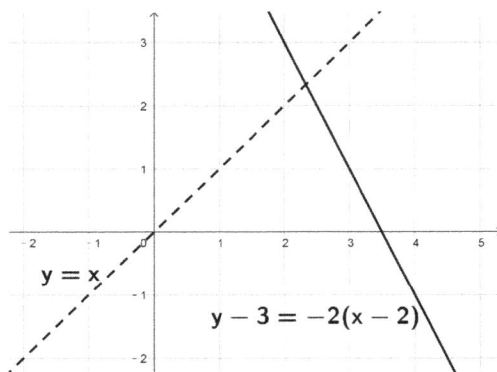

Note the second graph is shifted horizontally $+1$, shifted vertically -2, and is wider than the first graph.

Problem 5.2 Graph the following equations by plotting points.

(a) $y = x^3$.

Solution

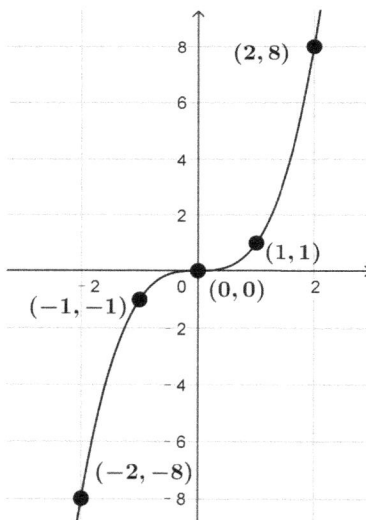

(b) $y = \sqrt{x}$

Solution

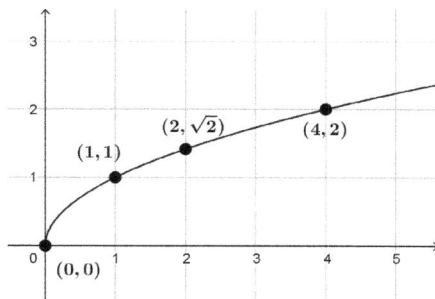

Problem 5.3 Graph the following equations.

(a) $y = 2x^3 - 6x^2 + 6x - 2$.

Note factoring we have $y = 2x^3 - 6x^2 + 6x - 2 = 2(x^3 - 3x + 3x - 1) = 2(x-1)^3$. This is the graph of $y = x^3$ shifted horizontally $+1$ and narrower than normal. The graph is shown below.

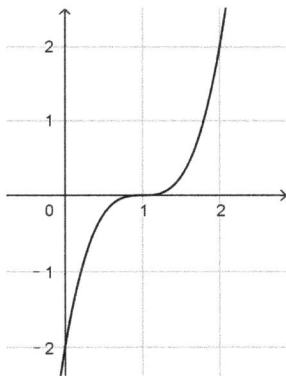

(b) $y = 2 - \sqrt{x - 2}$.

This is the graph of $y = \sqrt{x}$, except shifted horizontally $+2$ and vertically $+2$. Finally it is flipped upside-down. This gives the graph

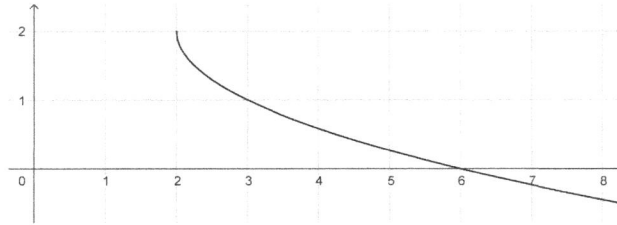

Problem 5.4 What is the domain and range of the following?

(a) $y + 3 = 2(x + 4)^2$.

> **Answer**

Domain: All real numbers, Range: All y with $y \geq -3$.

> **Solution**

This is a parabola opening upwards with vertex $(-4, -3)$. The domain is all real numbers and the range is all $y \geq -3$ (because the vertex $(-4, -3)$ is a minimum.

(b) $y - 4 = -(x - 2)^2/2$.

> **Answer**

Domain: All real numbers, Range: All y with $y \leq 4$.

> **Solution**

This is a parabola opening downwards with vertex $(2, 4)$. The domain is all real numbers and the range is all $y \leq 4$ (because the vertex $(2, 4)$ is a maximum.

Problem 5.5 Find the domain and range of the following functions.

(a) $f(x) = (x - 5)^5 + 5$.

> **Answer**

Domain: All real numbers, Range: All real numbers

Solution

The domain is all real numbers. Since we have an odd exponent, the range is also all real numbers.

(b) $f(x) = (x+2)^4 - 3$.

Answer

Domain: All real numbers, Range: All y with $y \geq -3$

Solution

This is the graph of $y = x^4$ shifted horizontally -2 and vertically -3. The domain is all real numbers. Since the exponent is even, the range is all $y \geq -3$.

(c) $f(x) = 3\sqrt{x-4} + 5$.

Answer

Domain: All x with $x \geq 4$, Range: All y with $y \geq 5$

Solution

The domain is all real numbers with $x \geq 4$, as we need $x - 4 \geq 0$ inside the square root. Square roots are always ≥ 0, so including the $+5$ the range is all $y \geq 5$.

Problem 5.6 Graph the following exponential functions by plotting points.

(a) $y = 2^x$.

Solution

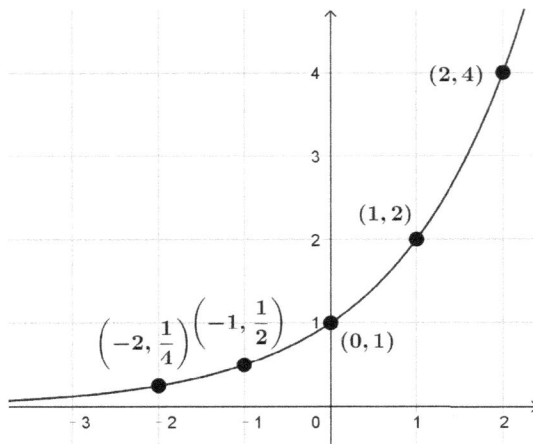

(b) $y = 3^x$.

Solution

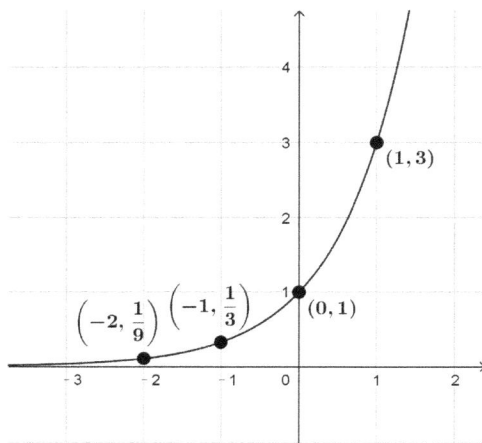

Problem 5.7 Exponential Function Graphing Continued

(a) Graph $y = \left(\dfrac{1}{2}\right)^x$. How does this compare to the graph of $y = 2^x$?

Solution

The graph is

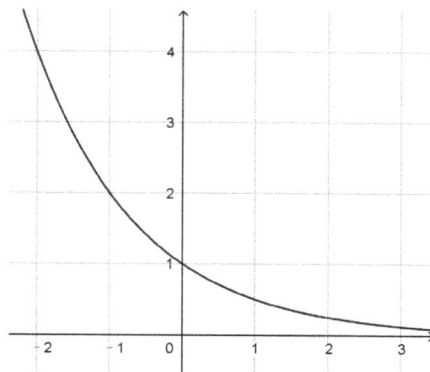

Note this is the same as the graph of $y = 2^x$ excepted flipped over the y-axis.

(b) Graph $y = \left(\dfrac{1}{3}\right)^x$. How does this compare to the graph of $y = 3^x$?

Solution

The graph is

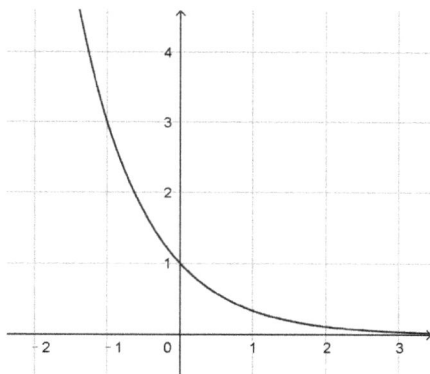

Note this is the same as the graph of $y = 3^x$ excepted flipped over the y-axis.

Problem 5.8 Exponents Revisited

(a) Calculate $\left(\dfrac{1}{3}\right)^{-3}$.

Answer

27

Solution

Note $\left(\dfrac{1}{3}\right)^{-3} = 3^3 = 27$.

(b) Calculate $8^{-2/3}$.

Answer

$\dfrac{1}{4}$

Solution

We have
$$8^{-2/3} = \frac{1}{8^{2/3}} = \frac{1}{\sqrt[3]{8}^2} = \frac{1}{2^2} = \frac{1}{4}.$$

(c) Calculate $9^{2/3}$ and express your answer in simplest radical form.

Answer

$3\sqrt[3]{3}$

Solution

We need to calculate $9^{2/3} = \sqrt[3]{9^2} = \sqrt[3]{81}$. Note $81 = 27 \times 3 = 3^3 \times 3$, so $\sqrt[3]{81} = 3\sqrt[3]{3}$.

Problem 5.9 Graph the following and give the domain and range of each.

(a) $y = 2 + 2^{x-2}$

Answer

Domain: All real numbers, Range: All y with $y \geq 2$.

Solution

The graph is that of $y = 2^x$ but shifted horizontally $+2$ and vertically $+2$:

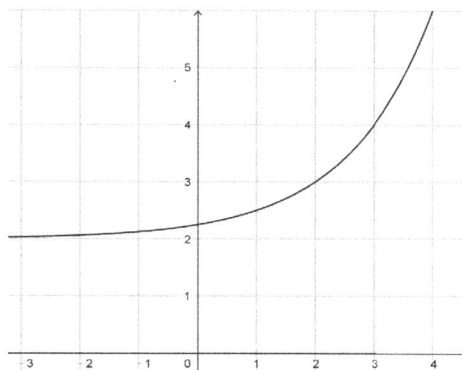

The domain is all real numbers x, with a range of all $y \geq 2$.

(b) $y = 3 \times \left(\dfrac{1}{3}\right)^x$

Answer

Domain: All real numbers, Range: All y with $y \geq 0$.

Solution

Note we have

$$y = 3 \times \left(\frac{1}{3}\right)^x = \left(\frac{1}{3}\right)^{-1} \times \left(\frac{1}{3}\right)^x = \left(\frac{1}{3}\right)^{x-1}$$

so the graph is that of $y = \left(\dfrac{1}{3}\right)^x$ shifted right $+1$:

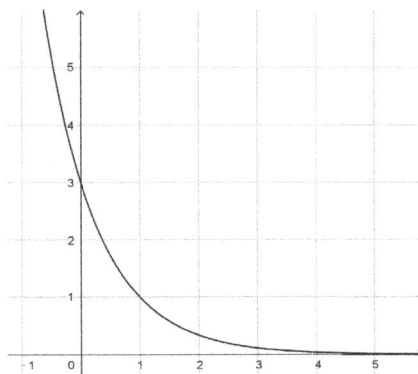

The domain is all real numbers x, with a range of all $y \geq 0$.

Problem 5.10 Suppose a certain type of bacterium divides once every hour. A certain petri dish starts with 4 bacteria. Write an equation representing the number of bacteria present after x hours and graph this equation.

Answer

$y = 2^{x+2}$

Solution

There are 4 bacteria at the start. Every hour after this the number of bacteria doubles (so the number is multiplied by 2). Hence the formula for the number of bacteria y after x hours is given by

$$y = 4 \times 2^x = 2^2 \times 2^x = 2^{x+2}.$$

This has graph

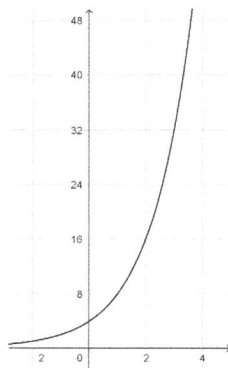

6 Solutions to Chapter 6 Examples

Problem 6.1 Review of Exponents

(a) Calculate $4^{3/2}$.

Answer

8

Solution

We have $4^{3/2} = \sqrt{4}^3 = 2^3 = 8$.

(b) Calculate $\left(\dfrac{2}{3}\right)^{-3}$

Answer

$\dfrac{27}{8}$

Solution

We have

$$\left(\frac{2}{3}\right)^{-3} = \left(\frac{3}{2}\right)^3 = \frac{27}{8}.$$

(c) Simplify $(\sqrt[4]{5})^{10} \times \dfrac{1}{5}$ by writing it in the form 5^A for some number A.

Answer

$5^{3/2}$

Solution

Working separately we have $(\sqrt[4]{5}^{10} = 5^{10/4} = 5^{5/2}$ and $\dfrac{1}{5} = 5^{-1}$. Hence $(\sqrt[4]{5})^{10} \times \dfrac{1}{5} = 5^{5/2} \times 5^{-1} = 5^{3/2}$.

Problem 6.2 Graph the equation $y = 4 - 3^{-x+2}$.

Solution

Note we can rewrite the equation as

$$y - 4 = -3^{-x+2} \Rightarrow y - 4 = -3^{-(x-2)} \Rightarrow y - 4 = -\left(\frac{1}{3}\right)^{x-2}.$$

So the graph of $y = 4 - 3^{-x+2}$ is the same as $y = \left(\frac{1}{3}\right)^x$ except: i) flipped upsidedown, ii) moved horizontally $+2$, iii) moved vertically $+4$. This gives the graph found below.

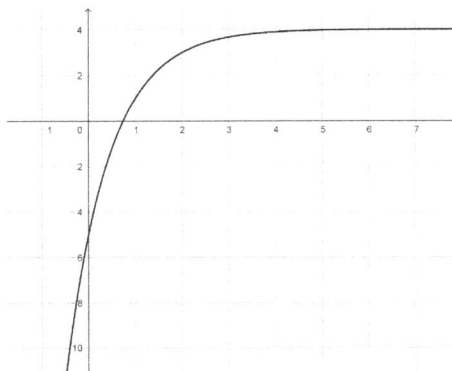

Problem 6.3 George starts with a stack of pennies. Each night before bed he makes a second stack with the same number of pennies as the first stack. Every morning me uses the two stacks to make one bigger stack of pennies. Suppose George starts with 3 pennies.

(a) If y denotes the number of pennies in George's stack, write an equation for y after x days.

Answer

$y = 3 \times 2^x$

Solution

At first there are 3 pennies. After one day, there are $3 + 3 = 6$ pennies, after 2 days there are $6 + 6 = 12$ pennies, etc. Noting the pattern the number of pennies doubles every day, starting at 3 pennies. This gives us the equation $y = 3 \times 2^x$.

(b) Equations of the form $y = C \times (1 + R)^x$ are examples of exponential growth. For your answer to part (a), explain the meaning of C and R.

Solution

Here $C = 3$ and $R = 1 = 100\%$. This means that the growth starts with $C = 3$ pennies, and increases by 100% (doubles) each day.

(c) How many pennies are in George's stack after 7 days?

Answer

384

Solution

After 7 days (when $x = 7$), George's stack will have $y = 3 \times 2^7 = 3 \times 128 = 384$ pennies.

Problem 6.4 Sara bought a new car, worth $\$15,000$. After some research, she determined that for each year she owned the car, the car would be worth 20% less than the year before.

(a) If y denotes the value of Sara's car in dollars, write an equation for y, after Sara owns the car for x years.

Answer

$y = 15000 \times 0.8^x$

Solution

Sara's car is initially worth 15000. Each year it is worth 20% less, so 80% of the previous year. Hence it is worth 15000×0.8 after one year, $15000 \times 0.8 \times 0.8$ after two years, etc. This gives an equation of $y = 15000 \times 0.8^x$.

(b) Equations of the form $y = C \times (1 - R)^x$ are examples of exponential decay. For your answer to part (a), explain the meaning of C and R.

Solution

$C = 15000$, the initial price of the car. $R = 0.2 = 20\%$, which is how much value the car's worth decreases each year.

(c) How much is Sara's car worth after 3 years?

Answer

$7680

Solution

After 3 years the car is worth $y = 15000 \times 0.8^3 = 15000 \times 0.512 = 7680$.

Problem 6.5 Peter is given $1000 by his grandfather to invest. Peter finds a good investment opportunity that will give him 10% returns per year.

(a) If he gets his return once a year, he earns 10% of his investment at the end of the year. If he gets his return twice a year, he earns $10\% \div 2 = 5\%$ of his investment after 6 months and another 5% at the end of the year. Does this change the amount he earns in one year?

Answer

No

Solution

Earning the 10% at the end of the year means Peter gets $10\% \times \$1000 = \$100$. Earning it as 5% twice he gets $2 \times 5\% \times \$1000 = \$100$ so it is the same either way.

(b) Suppose that whenever he earns money, Peter can reinvest that money. If Peter gets returns once a year, write an equation for how much money Peter's investment is worth after x years.

Answer

$y = 1000 \times (1.1)^x$

Solution

Peter gets a 10% return, meaning if he reinvests he has a total of $100\% + 10\% = 110\% = 1.1$ of the money he started with. Therefore after 1 year he has 1000×1.1 dollars, after 2 years he has $1000 \times (1.1)^2$ dollars, so after x years he has $y = 1000 \times (1.1)^x$ dollars.

(c) Suppose again that whenever he earns money, Peter can reinvest that money. If Peter gets returns twice a year, write an equation for how much money Peter's investment is worth after x years.

Answer

$y = 1000 \times (1.05)^{2x}$ or $y = 1000 \times (1.1025)^x$

Solution

Similar to part b), after each 5% return the total worth of his money is $100\% + 5\% = 105\% = 1.05$. Hence after 1 return he has 1000×1.05 dollars, after 2 returns he has 1000×1.05^2 dollars, etc. Since in this case Peter gets 2 returns per year, he has a total of $2x$ returns after x years. This gives an equation of $y = 1000 \times (1.05)^{2x}$. Note we can also rewrite this as $y = 1000 \times [(1.05)^2]^x = 1000 \times (1.1025)^x$.

(d) Which scenario is best for Peter? Explain your answer.

Answer

5% returns twice a year (compounded semianually)

Solution

Note in part b) we got an equation of $y = 1000 \times (1.1)^x$, where we see the rate $R = 1.1 - 1 = 0.1 = 10\%$. In part c) after simplifying we got $y = 1000 \times (1.1025)^x$ so the rate is $R = 1.1025 - 1 = 0.1025 = 10.25\%$ so we see that Peter has a slightly better rate when he gets returns twice a year.

Problem 6.6 Solve the following equations.

(a) $x^3 = 125$.

Answer

$x = 5$

Solution

Note that $125 = 5^3$ so we have $x^3 = 5^3$. Since they have the same exponent, the bases must be equal so $x = 5$.

(b) $2^x = 8$.

Answer

$x = 3$

Solution

Rewriting $8 = 2^3$ we must solve $2^x = 2^3$. Since both sides of the equation have the same base, the exponents must be equal so $x = 3$.

(c) $9^x = 243$.

Answer

$x = 5/2$

Solution

We notice that 243 is not a power of 9, but we can write $9 = 3^2$ and $243 = 5^2$. Hence we have $(3^2)^x = 3^5$ or $3^{2x} = 3^5$. Now that we have the same bases we see $2x = 5$ so $x = 5/2$.

Problem 6.7 Solve the following equations.

(a) $\frac{1}{5} \times 5^x = 125$

Answer

$x = 4$

Solution

Rewriting we have

$$\frac{1}{5} \times 5^x = 5^{-1} \times 5^x = 5^{x-1} = 125 = 5^3$$

so $x - 1 = 3$. Hence $x = 4$.

(b) $3 \left(\frac{1}{2}\right)^x = 192$

Answer

$x = -6$

Solution

Here we can instead divide by 3 on both sides of the equation to get

$$\left(\frac{1}{2}\right)^x = 2^{-x} = \frac{192}{3} = 64 = 2^6.$$

Therefore we have $-x = 6$ so $x = -6$.

Problem 6.8 Consider the data graphed below:

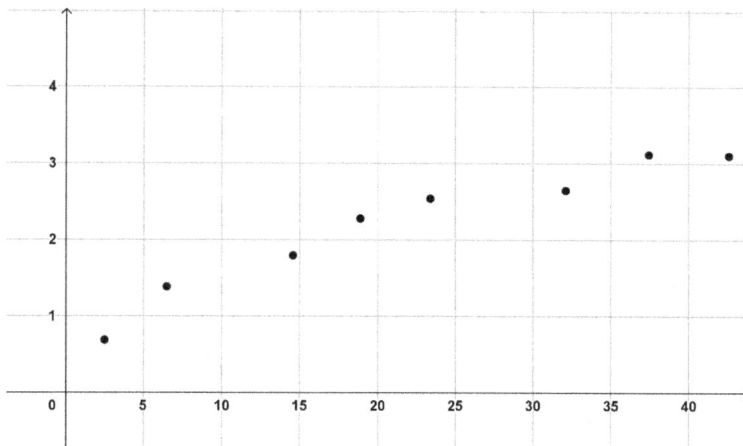

(a) If we want to model this data using one of the types of functions we have learned about (linear, quadratic, exponential, square root), which type of function do you think is most applicable? Why?

Answer

Square Root

Solution

The growth fairly slow, but is not linear. Hence modeling the data with a square root is the best choice.

(b) Using your answer from part (a), come up with an equation modeling the data.

Answer

$$y = \frac{1}{2}\sqrt{x}$$

Solution

Drawing a square root graph to fit the data gives the graph below.

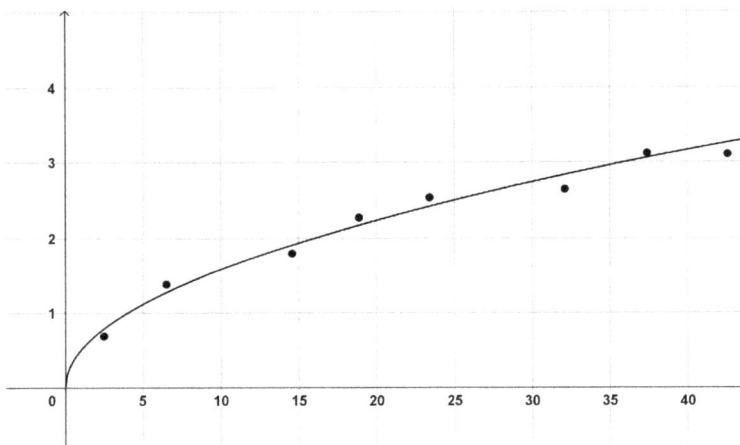

Look at square values for x, like $x = 4$, $x = 9$, $x = 16$, $x = 25$, etc. give us approximate points of

$$(4,1), (9,1.5), (16,2), (25,2.5).$$

Notice that that $1 = 2/1$, $1.5 = 3/2$, $2 = 4/2$, $2.5 = 5/2$, so we guess that the equation is $y = \frac{1}{2}\sqrt{x}$.

(c) Use your model to estimate the value of y when $x = 80$.

Answer

$y \approx 4.5$

Solution

Using the model of $y = \dfrac{1}{2}\sqrt{x}$, when $x = 80$ we have

$$y = \frac{1}{2}\sqrt{80} \approx \frac{1}{2}\sqrt{81} = \frac{1}{2} \times 9 = 4.5.$$

Problem 6.9 Consider the data in the following table

x	y
-2	0.11
-1	0.35
0	1.1
1	2.9
2	9.2

(a) If we want to model this data using one of the types of functions we have learned about (linear, quadratic, exponential, square root), which type of function do you think is most applicable? Why?

Answer

Exponential

Solution

The very fast growth leads to an exponential model.

(b) Using your answer from part (a), come up with an equation modeling the data.

Answer

$y = 3^x$

Solution

Note that the rounding the y values, we get the points $(0, 1), (1, 3), (2, 9)$. From here we might guess the equation $y = 3^x$ (as $3^0 = 1, 3^1 = 3, 3^2 = 9$). Double checking this makes sense, when $x = -1$, we have $y = 3^{-1} \approx 0.33$ and when $y = -2$, we have $y = 3^{-2} \approx 0.11$, so $y = 3^x$ is a good exponential model for our data.

(c) Use your model to estimate the value of x when $y = 735$.

Answer

$x \approx 6$

Solution

Using the model we want $y = 735$, or $735 = 3^x$. Note that 735 is not a power of 3, but is fairly close to $3^6 = 729$. Hence a good estimate for x will be $x = 6$.

Problem 6.10 Kim and Harry just launched their new app. They were excited to get 400 downloads by day 2 and 1600 downloads by day 4.

(a) Kim thinks the downloads are growing exponentially. Help Kim find a model for their downloads!

Answer

$y = 100 \times 2^x$

Solution

Kim thinks the model is exponential, so should be of the form $y = C \times (1 + R)^x$ or just $y = C \times A^x$ for some C and A. The model should contain the data $(2, 400)$ and $(4, 1600)$, so plugging these into the model we have $400 = C \times A^2$ and $1600 = C \times A^4$. Dividing these equations we have

$$\frac{1600}{400} = \frac{C \times A^4}{C \times A^2} \Rightarrow 4 = A^2 \Rightarrow A = \pm 2.$$

However since this is growth, we see that $A = 2$. Hence if we need $400 = C \times 2^2 = C \times 4$ we have $C = 100$. Therefore Kim's exponential model is $y = 100 \times 2^x$.

(b) Harry thinks their downloads are growing quadratically, with a vertex when $x = 0$. Help Harry find a model for their downloads!

Answer

$y = 100x^2$

Solution

Since Harry thinks the model is a quadratic (a parabola) with vertex when $x = 0$, we know the model should be of the form $y - k = a(x - 0)^2$ or $y = ax^2 + k$. The model should contain the data $(2, 400)$ and $(4, 1600)$, so plugging these into the model we have $400 = a \times 2^2 + k$ and $1600 = a \times 4^2 + k$. This gives us the system of equations $400 = 4a + k$ and $1600 = 16a + k$. Subtracting these two equations leads to

$$1600 - 400 = (16a + k) - (4a + k) \Rightarrow 1200 = 12a \Rightarrow a = 100.$$

From here we see $400 = 4 \times 100 + k$ so $k = 0$. Hence $y = 100x^2$ is the quadratic model.

(c) Are either Kim or Harry wrong? Discuss the pros and cons of each model.

Answer

Neither is necessarily wrong

Solution

Without more data to help decide, neither Kim and Harry are wrong. Kim's model assumes the growth will continue very rapidly, while Harry's model has slightly slower growth. In this way, neither model will be accurate in the long term, as the number of downloads must slow down eventually.

7 Solutions to Chapter 7 Examples

Problem 7.1 Solving simple equations involving square roots

(a) Solve $x^2 = 9$.

Answer

$x = \pm 3$

Solution

Taking the square root on both sides we have $x = \pm\sqrt{9} = \pm 3$. Do not forget that both $+\sqrt{9}$ and $-\sqrt{9}$ are solutions.

(b) Solve $\sqrt{x} = 3$.

Answer

$x = 9$

Solution

Squaring both sides we have $x = 9$.

(c) Solve $\sqrt{x} = -3$.

Answer

No solutions

Solution

Remember $\sqrt{x} \geq 0$, so there are no real solutions. We must be careful here as squaring both sides leads to $x = 9$, which does not work.

Problem 7.2 Solve the following equations.

(a) $2\sqrt{x} = 6 - \sqrt{x}$.

Answer

$x = 4$

Solution

Rewriting we have $3\sqrt{x} = 6$ so $\sqrt{x} = 2$. Hence squaring both sides we have $x = 4$.

(b) $\sqrt{x} + \sqrt{2x} = 1$.

Answer

$x = 3 - 2\sqrt{2}$

Solution

Recall that $\sqrt{2x}$ can be written as $\sqrt{2} \times \sqrt{x}$ so we have $\sqrt{x} + \sqrt{2} \times \sqrt{x} = \sqrt{x}(1 + \sqrt{2}) = 1$.
Therefore

$$\sqrt{x} = \frac{1}{1 + \sqrt{2}} = \frac{1}{1 + \sqrt{2}} \times \frac{1 - \sqrt{2}}{1 - \sqrt{2}} = \frac{1 - \sqrt{2}}{-1} = \sqrt{2} - 1.$$

Therefore $x = (\sqrt{2} - 1)^2 = 2 - 2\sqrt{2} + 1 = 3 - 2\sqrt{2}$.

Problem 7.3 Solve the following equations.

(a) $\sqrt{x - 3} + 4 = 8$.

Answer

$x = 19$.

Solution

We first have $\sqrt{x - 3} = 4$ so squaring both sides gives $x - 3 = 16$ and hence $x = 19$.

(b) $3 - \sqrt[3]{x + 5} = 8$.

Answer

$x = -130$

Solution

Isolating the cube root gives $-5 = \sqrt[3]{x+5}$ so cubing both sides gives $-125 = x+5$ so $x = -130$. Remember it is possible to have a cube root be negative.

Problem 7.4 Solve the following by squaring both sides. Be careful about extraneous roots!

(a) $\sqrt{x^2 + 3x} = 2$

Answer

$x = -4, 1$

Solution

Squaring both sides gives $x^2 + 3x = 4$ so $x^2 + 3x - 4 = (x+4)(x-1) = 0$. Hence $x = -4$ and $x = 1$ are solutions. Double check that both solutions work.

(b) Solve $\sqrt{2x^2 + 3x - 10} = x$.

Answer

$x = 2$

Solution

Squaring both sides we have $2x^2 + 3x - 10 = x^2$ so $x^2 + 3x - 10 = 0$. Factoring we have $(x+5)(x-2) = 0$ so $x = -5$ or $x = 2$. Note, however, if $x = -5$ we have $\sqrt{\ldots} = -5$ which is impossible. Hence $x = 2$ is the only solution (which double checking, works).

Problem 7.5 Solve the equation $\sqrt{1 + \sqrt{2+x}} = 3$.

Answer

$x = 62$

Solution

Squaring once gives $1 + \sqrt{2+x} = 9$ so $\sqrt{2+x} = 8$. Squaring again we have $2 + x = 64$. Hence we know that $x = 62$.

Problem 7.6 Solve the following equations by squaring twice.

(a) $\sqrt{x+3} - \sqrt{2-x} = 1$

Answer

$x = 1$

Solution

Isolating the $\sqrt{x+3}$ we have $\sqrt{x+3} = 1 + \sqrt{2-x}$. Squaring both sides we have $x+3 = 1 + 2\sqrt{2-x} + 2 - x$ so again isolating the square root gives $2x = 2\sqrt{2-x}$. Dividing by 2 and then squaring we have $x^2 = 2 - x$ so $x^2 + x - 2 = 0$. Factoring gives $(x+2)(x-1) = 0$ so $x = -2$ or $x = 1$. Note $x = -2$ does not work, but $x = 1$ does.

(b) $\sqrt{2x} - \sqrt{2x-10} = 4$.

Answer

No solutions

Solution

We have $\sqrt{2x} = \sqrt{2x-10} + 8$. Squaring we have $2x = 2x - 10 + 8\sqrt{2x-10} + 16$ so $-6 = 8\sqrt{2x-10}$. Note this is impossible, hence there are no solutions.

Problem 7.7 Recall that $x^{a/b} = (\sqrt[b]{x})^a$. Solve the following equations for x.

(a) $x^{-1/3} = \dfrac{1}{3}$

Answer

$x = 27$

Solution

We have
$$x^{-1/3} = \frac{1}{3} \Rightarrow \frac{1}{x^{1/3}} = \frac{1}{3} \Rightarrow \sqrt[3]{x} = 3 \Rightarrow x = 27.$$

(b) $(r-1)^{-2/3} = \dfrac{1}{4}$.

Answer

$x = -7, 9$

Solution

We have $(r-1)^{-2/3} = \dfrac{1}{(r-1)^{2/3}}$ so after cross multiplying we want to solve $(r-1)^{2/3} = \sqrt[3]{(r-1)^2} = 4$. Cubing both sides gives $(r-1)^2 = 64$ and thus $r-1 = \pm 8$. Hence $r = 9$ and $r = -7$.

Problem 7.8 Solve the equation $\sqrt{|x| + x} = 4$.

Answer

$x = 8$

Solution

If $x < 0$, $|x| + x = -x + x = 0$, so $\sqrt{|x| + x} = 0 \neq 4$. For $x \geq 0$, $|x| + x = 2x$ so we want $\sqrt{2x} = 4$. Squaring gives $2x = 16$ so $x = 8$.

Problem 7.9 Jenna and Olivia both try to model the growth of their basil plant. Jenna models the growth using the equation $y = x + 1$ while Olivia models using the equation $y = \sqrt{x^2 - 2x + 3}$.

(a) Do their models ever agree?

Answer

Yes, at $\left(\dfrac{1}{2}, \dfrac{3}{2} \right)$

Solution

The models agree if $x + 1 = \sqrt{x^2 - 2x + 3}$. Squaring both sides we have $x^2 + 2x + 1 = x^2 - 2x + 3$. Collecting like terms we have $4x = 2$ so $x = \dfrac{1}{2}$. Plugging in $x = \dfrac{1}{2}$ we get $y = \dfrac{1}{2} + 1 = \dfrac{3}{2}$.

(b) For what x is Jenna's predicted growth larger than Olivia's model?

Answer

$x > \dfrac{1}{2}$

Solution

Note in part a) we solve to see that the models only agree when $x = \dfrac{1}{2}$. Thus Jenna's predicted growth is larger either when $x < \dfrac{1}{2}$ or when $x > \dfrac{1}{2}$. Checking $x = 1$ Jenna's model has $y = 1 + 1 = 2$ while Olivia's has $y = \sqrt{1^2 - 2 \times 1 + 3} = \sqrt{2}$. Since $2 > \sqrt{2}$, we see Jenna's predicted growth is larger for all $x \geq \dfrac{1}{2}$.

Problem 7.10 Find the domain and range of $y = \sqrt{x^2 - 4x + 8}$.

Answer

Domain: All real numbers. Range: $y \geq 2$.

Solution

Completing the square we have $x^2 - 4x + 8 = x^2 - 4x + 4 - 4 + 8 = (x-2)^2 + 4$ so we have $y = \sqrt{(x-2)^2 + 4}$. Note the inside of the square root the equation of a parabola with vertex $(2, 4)$. Therefore the domain is all real numbers, and as $(x-2)^2 + 4$ has a minimum value of 4, the minimum value of $\sqrt{(x-2)^2 + 4}$ is $\sqrt{4} = 2$. Thus the range is all $y \geq 2$.

8   Solutions to Chapter 8 Examples

Problem 8.1 Expand the following.

(a) $(t-2)(t^2-t+3)$

Answer

t^3-3t^2+5t-6

Solution

Distributing we have

$$t^3-t^2+3t-2t^2+2t-6=t^3-3t^2+5t-6.$$

(b) $(x^3+2x+3)(x^2+4x+2)$

Answer

$(x^5+4x^4+4x^3+11x^2+16x+6)$

Solution

Distributing we have

$$x^5+4x^4+2x^3+2x^3+8x^2+4x+3x^2+12x+6=x^5+4x^4+4x^3+11x^2+16x+6.$$

Problem 8.2 Review of Long Division, Quotients, and Remainders

(a) Calculate $789\div 6$ using long division.

Answer

131.5

(b) Write $789=Q\times 6+R$ where Q is the quotient and R is the remainder when dividing 789 by 6.

Answer

$789 = 131 \times 6 + 3$

(c) Write $122 = Q \times 4 + R$ where Q is the quotient and R is the remainder when dividing 122 by 4.

Answer

$122 = 30 \times 4 + 2$

Problem 8.3 Division with Polynomials

(a) Divide $(a - 4)$ by a.

Answer

$1 - \dfrac{4}{a}$

Solution

Note dividing by a is the same as multiplying by $\dfrac{1}{a}$. Hence distributing we have

$$\frac{a-4}{a} = \frac{1}{a}(a-4) = 1 - \frac{4}{a}.$$

(b) Verify that $\dfrac{x^2}{x-2} = (x+2) + \dfrac{4}{x-2}$.

Solution

Multiplying both sides of the equation by $x - 2$ to clear the denominator we have $x^2 = (x+2)(x-2) + 4$ which is true because $(x+2)(x-2) + 4 = x^2 - 4 + 4 = x^2$.

Problem 8.4 Calculate the quotient and remainder of $(x^3 + 2) \div (x + 1)$.

Answer

Quotient: $x^2 - x + 1$, Remainder: 1

Solution

Polynomial long division gives us $x^3 + 2 = (x^2 - x + 1) \times (x + 1) + 1$ so the quotient is $x^2 - x + 1$ with remainder 1.

Problem 8.5 Calculate $(2v^4 + v^3 - 6v^2 - v + 1) \div (2v + 1)$.

Answer

$v^3 - 3v + 1$

Solution

Polynomial long division gives us $2v^4 + v^3 - 6v^2 - v + 1 = (v^3 - 3v + 1) \times (2v + 1)$ so

$$\frac{2v^4 + v^3 - 6v^2 - v + 1}{2v + 1} = v^3 - 3v + 1.$$

Problem 8.6 Calculate $(x^3 + 2x^2 + x + 1) \div (x^2 + x + 1)$.

Answer

$x + 1 - \dfrac{x}{x^2 + x + 1}$

Solution

Polynomial long division gives us that $x^3 + 2x^2 + x + 1 = (x + 1) \times (x^2 + x + 1) - x$ so

$$\frac{x^3 + 2x^2 + x + 1}{x^2 + x + 1} = x + 1 - \frac{x}{x^2 + x + 1}.$$

Problem 8.7 What is the domain of the following functions?

(a) $y = \dfrac{x^2 - 2x - 3}{x - 1}$

Answer

All $x \neq 1$

Solution

The function is defined as long as $x - 1 \neq 0$. Hence the domain is all x such that $x \neq 1$.

(b) $y = \dfrac{1}{x^2 + x + 1}$

Answer

All x

Solution

The function is defined as long as $x^2 + x + 1 \neq 0$. Since the discriminant of $x^2 + x + 1$ is $1^2 - 4 \times 1 \times 1 = -3$ the equation $x^2 + x + 1 = 0$ has no roots. Therefore the domain is all real numbers x.

Problem 8.8 Graph the equation $y = \dfrac{x^2 + 2x - 3}{x - 1}$.

Solution

Note factoring we have

$$y = \frac{x^2 + 2x - 3}{x - 1} = \frac{(x - 1)(x + 3)}{x - 1} = x + 3 \text{ when } x \neq 1.$$

Hence the graph is just the line $y = x + 3$ except undefined when $x = 1$. This is shown below:

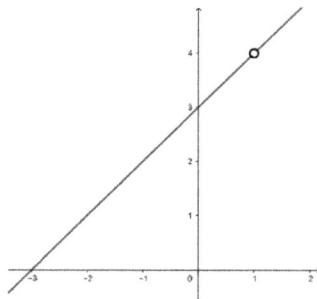

Problem 8.9 Graph the equation $y = \dfrac{x^2 + 2x + 4}{x^2 - 4}$. What are the asymptotes?

Answer

Vertical: $x = \pm 2$, Horizontal: $y = 1$

Solution

Using long division we have

$$y = 1 + \frac{2x + 8}{x^2 - 4} = 1 + \frac{2x + 8}{(x - 2)(x + 2)}.$$

Therefore the graph has vertical asymptotes of ± 2 and a slant asymptote of $y = 1$. The full graph is shown below

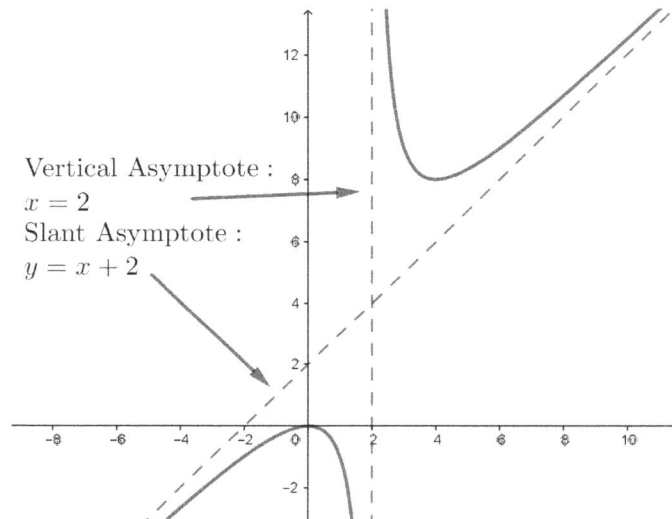

Note this full graph is quite complicated, but you should at least understand the asymptotes at this point.

Problem 8.10 Graph the equation $y = \dfrac{x^2}{x - 2}$. What are the asymptotes?

Answer

Vertical: $x = 2$, Slant: $y = x + 2$

Solution

Using long division we have

$$y = x + 2 + \frac{4}{x-2}.$$

Therefore the graph has a vertical asymptote of 2 and a slant asymptote of $y = x + 2$. The full graph is shown below:

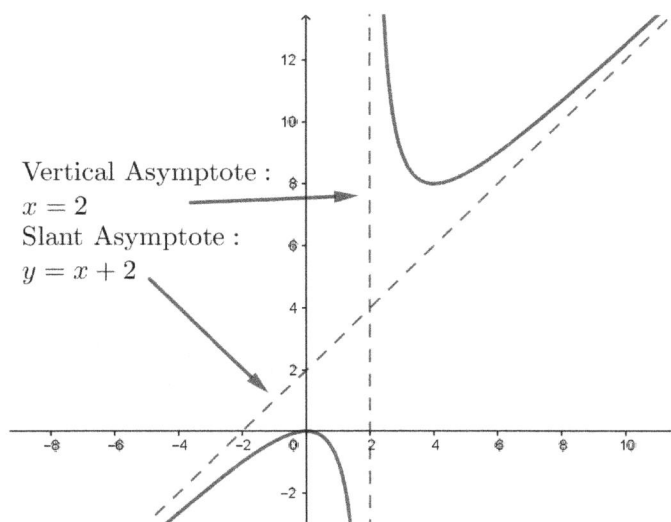

Vertical Asymptote :
$x = 2$
Slant Asymptote :
$y = x + 2$

9   Solutions to Chapter 9 Examples

Problem 9.1   What is the quotient and remainder when $x^5 + 3x^2 - 2x + 1$ is divided by $x^2 + x + 1$?

Answer

Quotient: $x^3 - x^2 + 4$, Remainder: $-6x - 3$

Solution

Using polynomial long division we calculate a quotient of $x^3 - x^2 + 4$ with remainder $-6x - 3$. Double checking, note

$$(x^2 + x + 1)(x^3 - x^2 + 4) - 6x - 3 = x^5 + 3x^2 - 2x + 1.$$

Problem 9.2   Verify the Polynomial Remainder Theorem for the following. That is show that the remainder of $P(x) \div (x - a)$ is equal to $P(a)$.

(a) $P(x) = x^3 + x^2 + x + 6$ and $a = -2$.

Answer

$r = P(-2) = 0$

Solution

We have $P(-2) = (-2)^3 + (-2)^2 - 2 + 6 = -8 + 4 - 2 + 6 = 0$. Double checking, we have

$$(x^3 + x^2 + x + 6) \div (x + 2) = x^2 - x + 3$$

so there is no remainder.

(b) $P(x) = x^3 + 2x^2 + x - 3$ and $a = 5$.

Answer

$r = P(5) = 177$

Solution

We have $P(5) = 5^3 + 2 \times 5^2 + 5 - 3 = 125 + 50 + 5 - 3 = 177$. Double checking, we have

$$(x^3 + 2x^2 + x - 3) \div (x - 5) = x^2 + 7x + 36 + \frac{177}{x - 5}$$

so the remainder is 177.

Problem 9.3 Solve the following equations using the given hint.

(a) $x^3 - 5x^2 + 5x + 3 = 0$ if $x = 3$ is one solution.

Answer

$x = 3, x = 1 \pm \sqrt{2}$

Solution

We know $x = 3$ is a solution, so $(x - 3)$ is a factor of $x^3 - 5x^2 + 5x + 3$. Using long division we have

$$x^3 - 5x^2 + 5x + 3 = (x - 3)(x^2 - 2x - 1) = 0.$$

Using the quadratic formula $x^2 - 2x - 1 = 0$ has roots

$$\frac{2 \pm \sqrt{4 - 4 \times -1}}{2} = \frac{2 \pm 2\sqrt{2}}{2} = 1 \pm \sqrt{2}.$$

Hence the three solutions are $x = 3$, $x = 1 \pm \sqrt{2}$.

(b) $x^4 + x^3 - 6x^2 - 4x + 8 = 0$ if $x = 1$ and $x = -2$ are solutions.

Answer

$x = 1, x = \pm 2$

Solution

We know $x = 1$ and $x = -2$ are solutions, so $(x - 1)(x + 2) = x^2 + x - 2$ is a factor of $x^4 + x^3 - 6x^2 - 4x + 8$. Long division gives us

$$x^4 + x^3 - 6x^2 - 4x + 8 = (x^2 + x - 2)(x^2 - 4) = (x - 1)(x + 2)(x - 2)(x + 2) = 0.$$

Hence the solutions are $x = 1, x = \pm 2$. (Here $x = -2$ is a double root.)

Problem 9.4 Suppose the remainder when $(kx^2 - x + 2) \div (x - 5)$ is -53. What is k?

Answer

$k = -2$

Solution

The remainder when divided by $x - 5$ is -53, which is the same by the Polynomial Remainder Theorem as $kx^2 - x + 2$ when $x = 5$. Hence

$$k \times 5^2 - 5 + 2 = -53 \Rightarrow 25k = -50 \Rightarrow k = -2.$$

Problem 9.5 Consider $P(x) = 2x^3 - ax^2 - bx - 1$. Suppose the remainder when $P(x) \div (x + 1)$ is -10 and the remainder when $P(x) \div (x + 2) = -37$. What are a and b?

Answer

$a = 3, b = -4$

Solution

We use the Polynomial Remainder Theorem. This tell us that $P(-1) = -10$ and $P(-2) = -37$. Hence

$$2(-1)^3 - a(-1)^2 - b(-1) - 1 = -10 \Rightarrow b - a = -7$$

and similarly

$$2(-2)^3 - a(-2)^2 - b(-2) - 1 = -37 \Rightarrow 2b - 4a = -20.$$

Doubling the first equation gives $2b - 2a = -14$ and subtracting this from the second equation we have $-2a = -6$ so $a = 3$. Thus $b - 3 = -7$ so $b = -4$.

Problem 9.6 Practice with the Rational Root Theorem

(a) What are the possible rational roots of $2x^3 - 11x^2 + 4x + 5 = 0$?

Answer

$\pm 1, \pm 5, \pm 0.5, \pm 2.5$

Solution

The rational root theorem says that a rational root much have numerator a factor of ± 5 and denominator a factor of ± 2. This gives the possiblities

$$\pm\frac{1}{1}, \pm\frac{5}{1}, \pm\frac{1}{2}, \pm\frac{5}{2},$$

or $\pm 1, \pm 5, \pm 0.5, \pm 2.5$.

(b) Use the possible rational roots to help solve equation $2x^3 - 11x^2 + 4x + 5 = 0$.

Answer

$x = 1, x = -0.5, x = 5$

Solution

Trying the rational roots from a) we see that $x = 1$ is a root. Hence $x - 1$ is a factor of $2x^3 - 11x^2 + 4x + 5$. Long division gives

$$2x^3 - 11x^2 + 4x + 5 = (x - 1)(2x^2 - 9x - 5) = (x - 1)(2x + 1)(x - 5) = 0.$$

Hence the roots are $x = 1$, $x = -0.5$, and $x = 5$.

Alternatively it is possible in this case to use the Rational Root Theorem to guess all 3 roots.

Problem 9.7 Write each of the following rational functions in the form $y = \dfrac{P(x)}{Q(x)}$ for polynomials $P(x)$ and $Q(x)$, simplifying where possible. What is the domain?

(a) $y = \dfrac{1}{x^2 + x} + \dfrac{x^2}{x + 1}$.

Answer

$y = \dfrac{x^3 + 1}{x^2 + x}$, Domain: All x with $x \neq 0, 1$

Solution

Factoring the denominators we have $y = \dfrac{1}{x(x + 1)} + \dfrac{x^2}{x + 1}$. Thus a common denominator is $x(x+1)$ and we have

$$y = \frac{1}{x(x+1)} + \frac{x(x^2)}{x(x+1)} = \frac{x^3+1}{x^2+x}.$$

Note in fact we can factor further, as $x^3+1 = (x+1)(x^2-x+1)$ so we have $y = \dfrac{x^2-x+1}{x}$ after cancelling, but this is not the focus of this chapter.

(b) $y = \dfrac{x}{x+1} - \dfrac{2}{x+2}.$

Answer

$y = \dfrac{x^2-2}{(x+1)(x+2)}$, Domain: All x with $x \neq -1, -2$

Solution

We see the common denominator is $(x+1)(x+2)$. Hence

$$y = \frac{x(x+2)}{(x+1)(x+2)} - \frac{2(x+1)}{(x+2)(x+1)} = \frac{x^2+2x-2x-2}{(x+1)(x+2)} = \frac{x^2-2}{(x+1)(x+2)}.$$

Problem 9.8 Solve the equation $\dfrac{1}{x} + \dfrac{1}{x+1} + \dfrac{1}{x-1} = 0.$

Answer

$x = \pm\dfrac{\sqrt{3}}{3}$

Solution

We know that x cannot be 0 or ± 1. Multiplying both sides of the equation by $x(x-1)(x+1)$ we have

$$(x-1)(x+1) + x(x-1) + x(x+1) = x^2-1+x^2-x+x^2+x = 0$$

so we have $3x^2 = 1$ and thus $x = \pm\dfrac{1}{\sqrt{3}} = \pm\dfrac{\sqrt{3}}{3}.$

Problem 9.9 Solve the equation $\dfrac{2}{x^2+3x+2} = 1 - \dfrac{x}{x^2-1}.$

Answer

$$x = 0, x = -\frac{1}{2} \pm \frac{\sqrt{21}}{2}$$

Solution

Factoring the denominators, note $x^2 + 3x + 2 = (x+1)(x+2)$ and $x^2 - 1 = (x-1)(x+1)$. Hence x cannot be ± 1 or -2. Multiplying by $(x+1)(x+2)(x-1)$ we have

$$2(x-1) = (x+1)(x+2)(x-1) - x(x+2) \Rightarrow 2x - 2 = x^3 + 2x^2 - x - 2 - x^2 - 2x.$$

Therefore after simplifying we have

$$x^3 + x^2 - 5x = x(x^2 + x - 5) = 0.$$

Thus $x = 0$ is one solution, and using the quadratic formula the other solutions are

$$x = \frac{-1 \pm \sqrt{1 - 4 \times -5}}{2} = \frac{-1}{2} \pm \frac{\sqrt{21}}{2}.$$

Problem 9.10 Suppose you expand $(x-1)^7$ to get a polynomial (with degree 7). What is the sum of all the coefficients of this polynomial?

Answer

0

Solution

If we expand $(x-1)^7$ we would get

$$(x-1)^7 = Ax^7 + Bx^6 + Cx^5 + \cdots + Gx + H.$$

Note then if we set $x = 1$ we have

$$(1-1)^7 = A(1)^7 + B(1)^6 + C(1)^5 + \cdots + G(1) + H,$$

so in fact

$$0 = A + B + C + \cdots + G + H$$

and the sum of all the coefficients of this polynomial is 0.

Alternatively you can use Pascal's triangle to help actually calculate each of the coefficients.

www.ingramcontent.com/pod-product-compliance
Lightning Source LLC
Chambersburg PA
CBHW081505200326
41518CB00015B/2392

GREYSCALE

BIN TRAVELER FORM

Cut By _Bertha Ramirez_   Qty _102_   Date _3-30-26_
       #18

Scanned By _____ Qty _____ Date _____

Scanned Batch ID's

_____ _____ _____

Notes / Exception